Sabermetrics
Baseball, Steroids, and How the Game has Changed Over the Past Two Generations

Sabermetrics

Baseball, Steroids, and How the Game has Changed Over the Past Two Generations

Gabriel B. Costa

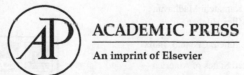

ACADEMIC PRESS

An imprint of Elsevier

Academic Press is an imprint of Elsevier
125 London Wall, London EC2Y 5AS, United Kingdom
525 B Street, Suite 1650, San Diego, CA 92101, United States
50 Hampshire Street, 5th Floor, Cambridge, MA 02139, United States
The Boulevard, Langford Lane, Kidlington, Oxford OX5 1GB, United Kingdom

Notices
Knowledge and best practice in this field are constantly changing. As new research and
experience broaden our understanding, changes in research methods, professional practices,
or medical treatment may become necessary.

Practitioners and researchers must always rely on their own experience and knowledge in
evaluating and using any information, methods, compounds, or experiments described herein.
In using such information or methods they should be mindful of their own safety and the
safety of others, including parties for whom they have a professional responsibility.

To the fullest extent of the law, neither the Publisher nor the authors, contributors, or editors,
assume any liability for any injury and/or damage to persons or property as a matter of
products liability, negligence or otherwise, or from any use or operation of any methods,
products, instructions, or ideas contained in the material herein.

British Library Cataloguing-in-Publication Data
A catalogue record for this book is available from the British Library.

Library of Congress Cataloging-in-Publication Data
A catalog record for this book is available from the Library of Congress.

ISBN: 978-0-12-822345-1

For Information on all Academic Press publications
visit our website at https://www.elsevier.com/books-and-journals

Publisher/Acquisitions Editor: Katey Birtcher
Editorial Project Manager: Naomi Robertson
Publishing Services Manager: Shereen Jameel
Project Manager: Kamatchi Madhavan
Cover Designer: Miles Hitchen

Typeset by MPS Limited, Chennai, India

Printed in the United States of America

Last digit is the print number: 9 8 7 6 5 4 3 2 1

Working together
to grow libraries in
developing countries

www.elsevier.com • www.bookaid.org

To the memory of Dr. Charles L. Suffel, who taught Mathematics at Stevens Institute of Technology for a half century, and whose friendship was a blessing to all…

Table of Contents

Preface

Hello. My name is Gabe Costa. I am a Catholic priest and a Mathematics professor. I am also a baseball fan and have been rooting for my beloved New York Yankees for nearly 65 years.

The two interests of mathematics and baseball would ultimately converge into what would become known as *Sabermetrics*.

My oldest friend, Frank Mottola, whom I met in kindergarten in 1953, introduced me to early issues of Bill James' ***Baseball Abstract*** in the 1980s. After I read these seminal books, I mulled over the possibility of creating a university-credited course on sabermetrics.

With the support and assistance of three wonderful educators (Dr Dan Gross, one of the contributors to this book; the late Dr John J Saccoman, the father of another one of the contributors; and the late Dr Jane Norton), Seton Hall University approved MATH 1011 – Sabermetrics. The course, which was first offered in January of 1988, is still running at Seton Hall (which, incidentally, has a tremendous baseball program) and is believed to be the first ever course of its kind.

Since then, many schools have created similar courses, including the United States Military Academy here at West Point, New York.

At this writing, sabermetrics (which Bill James defined as *the search for objective truth about baseball*, taking the "saber" part from SABR, the acronym for the Society for American Baseball Research) is in its third generation. In fact, in many contexts, the word has been replaced by *analytics*. And virtually every team sport and individual sport has a type of "sports analytics" associated with it.

Over the past 50 years or so, many books have been written about sabermetrics (e.g., contributors Mike Huber and John T Saccoman, along with yours truly, have published four volumes). Some of these books can be used as texts in the classroom, such as ***Curve Ball*** by Jim Albert. John Thorn and Pete Palmer published a classic book titled ***The Hidden Game of Baseball***. Robert Adair penned the timeless ***The Physics of Baseball***. Baseball historian, Bill Jenkinson, has researched home runs for nearly 50 years, has written innumerable articles on the topic, and has published at least two books.

Of course, Baseball is more than numbers. "Whoever wants to know the heart and mind of America had better learn baseball…" is a well-known adage authored by historian Jacques Barzun. Surely the game is "easily" associated with such other topics as:

- Baseball and Americana
- Baseball and Economics
- Baseball and the Movies

- Baseball and Race Relations
- Baseball and Steroids
- Baseball and the World Wars

In our book, the reader will be treated to a number of essays dealing with a myriad of subjects. Some articles will "overlap" with others; some will seem to be "at odds" with others; still others will put forth statements that I consider as bordering on baseball heresy (as Mike Scioletti—a former student, no less—asserts about Babe Ruth!).

The only prerequisite for the reader is to have a *passion* for baseball. Some basic knowledge of algebra and statistics will assist the reader, but it is hardly necessary for all but a few pages of this volume.

Our aim is to give the reader a book that can be read and re-read; one where he/she can open the text anywhere and then commence reading.

It is also hoped that fans of the National Pastime will enjoy this book, whether they are reading it from perspectives that might be historical, sociological, economical, or statistical (see above).

There are nearly a dozen authors who have contributed to this collection. Each writer has his own styles, and I have asked the publisher to preserve these styles as much as possible. Where statistics, and other baseball information included, are without a reference, it has likely been taken from the website: https://www.baseball-reference.com/ The reader will notice that some authors end their pieces with discussion questions; others do not. It would be hard to pick a better team of collaborators; so gentlemen, I thank you for your contributions!

Before closing, I want to thank all the people at Elsevier who assisted with this project, especially Ms Katey Birtcher, Ms Naomi Robertson (who is across the Pond in England), and Ms Kamatchi Madhavan (who is across the *other* Pond in India).

Lastly, I thank the Lord for creating such a perfect game. As the greatest of all players penned, a few months before his death:

"For above everything else, I want to be part of and help the development of the greatest game God ever saw fit to let man invent – baseball."
Babe Ruth (1948)

Gabriel B. Costa
West Point, New York

About the Contributors

Chris Arney is a retired Army Brigadier General and holds a Ph.D. in Mathematics. He splits rooting interests between the Atlanta Braves and the New York Mets.

Gabriel B. Costa is a Catholic priest on an extended academic leave from Seton Hall University. He holds a Ph.D. in Mathematics and presently teaches Mathematics and serves as an Associate Chaplain at West Point. He has been a Yankee fan since 1958.

John DeSomma, MBA, is a retired Industrial Engineer. He has been a New York Yankees fan since the mid-1950s.

Lee Evans is a Lieutenant Colonel in the United States Army. He holds a Ph.D. in Industrial Engineering and teaches Mathematics at West Point. He is a Houston Astros fan by birth, and a Chicago Cubs fan by marriage.

Jesse Germain is a retired Army Colonel. He holds a Ph.D. in Health and Human Performance and is presently the Director of Kinesiology in the Department of Physical Education at West Point. He roots for the New York Mets.

Daniel Gross holds a Ph.D. in Mathematics and teaches at Seton Hall University. He and his wife Mary are life-long fans of the New York Yankees.

Michael Huber is a retired Army Lieutenant Colonel. He holds a Ph.D. in Mathematics and teaches at Muhlenberg College. He roots for the Baltimore Orioles.

Eugene Reynolds holds two Master's degrees, one in Humanities and one in Environmental Management. He presently teaches Mathematics at Seton Hall University and describes himself as an "ecumenical baseball fan."

John T. Saccoman, Ph.D., is Professor and Chair of the Department of Mathematics and Computer Science at Seton Hall University. His favorite team is the New York Mets.

Michael Scioletti is a Colonel in the United States Army and is the Deputy Head of the Department of Mathematical Sciences at West Point. He holds a Ph.D. in Operations Research and is a zealous Boston Red Sox fan.

An appreciation of baseball and its mathematics

1

Chapter outline

What is baseball? The answer depends on how complete, complex, and methodical you want to consider the many faces and perspectives of the game. Perhaps a full definition of baseball lies within the entirety of this book. For many baseball fans, the essence of baseball lies in their heart and soul. And for others, the nature of baseball reveals itself in numbers and statistics within their mind.

In this overture, I will try to hit only a few high notes by confining this section to some of the most obvious elements of baseball and mostly at the major league level. I will try to describe baseball in 10 parts as (in order) baseball's popularity, a feature of my own heart and mind, an amazing sport with plentiful data, a game with wonderful statistics, a business with huge egos and desires for profits, a fulfilling form of entertainment, recreational fun, a developer of teamwork and fitness in youth, a major cultural influencer, and for over a century America's national pastime. My own perspectives stem from being both a fan and a lover of baseball's numbers, metrics, and models. Throughout these discussions, I will offer some of baseball's fascinating history. And to conclude the overture, I will try to envision the future roles of the game and give you encouragement through a few problems to consider. This entire book and each of its chapters invite readers to go deeper into the topics covered and way beyond the elements and ideas in this overture.

Sabermetrics. DOI:0.1016/B978-0-12-822345-1.00001-5

Baseball's popularity

Let's first consider the question of why people become baseball fans, or maybe in a more general form, why do fans love baseball so much? Sometimes the intense feeling of the fan-to-sport connection is rational. The fan may enjoy the intricate elements of the game, such as the strategy of the coaching, the pace of the game so that nuances can be considered, the myriad statistics, the physics of the ball in the air and the movement of the bat, the geometry of the diamond and the location of the defensive players, or the athleticism and actions of the players. However, these kinds of fans are few, and it is rare for fans to look at baseball as a subject in school or a conventional profession to be admired. There are many more fans that get hooked through an emotional passion for the competitive, natural, and exciting elements of the game—the legacy of a hometown team, a family tradition, the thrill of the game (especially those walk-off home runs), the feeling of connection to a winning team or a form of extra support for a perennial underdog, the entertainment and fun at the stadium, or the connections to the players as heroes for the community. Most fans have a unique combination of some or all of these parts of the game that stoke their fanatical feelings for the wonders and manifestations of baseball.

Henry Chadwick started developing and refining basic baseball statistics in the 19th century. He developed box scores and baseball guides with statistical summaries. His work was so significant some call him the "Father of Baseball," and he was elected to the Hall of Fame in 1938. The more modern form of sabermetrics was slow to develop, and it stayed hidden from fans for a while. Its pioneers, such as Bill James, used their more advanced metrics to help teams find the best players, and eventually the power of more accurate measurements for the most popular game in America came to light. The Society for American Baseball Research (SABR) was established in 1971 and fans started to catch on that the old simple ratios on the backside of baseball cards were inadequate to determine real quality of performance. sabermetrics concentrated on the various ways a player's performance produced runs or wins and what value a player brought to the team's wins over an (average) replacement player. Later, Nate Silver developed algorithms to compare and forecast the trajectory of a player's career to the performance of similar players. This book will explain how this form of applied statistics and modeling works well for the game of baseball.

In a somewhat related and simultaneous movement, fantasy baseball arrived on the scene and captured a new form of interest in baseball. The fantasy game of playing the roles of team owner, general manager (GM), and manager started with simple tabletop games and rotisserie systems to pick players. The play can be head-to-head with day-to-day standings or a cumulative competition using several players' or the team's measures of performance (sometimes sabermetric). Online, real-time Internet systems make the fantasy form of the games seems live and exciting for every day of the season. Nowadays, fantasy leagues have commissioners, fees and prizes, trades, budgets, salaries, and year-to-year rollovers of players and team owners.

Some fans develop their connection to baseball early in their life, like a child-like infatuation. This happens when the child plays the sport in Little League or plays softball in a school or recreation league. Others come to the game much later as adults with a more profound form of attachment even if it is strictly as spectators or fantasy players. Some sustain their ardent feelings for their lifetime or for decades, while they grow and change even as the sport itself also grows and changes. Other fair-weather fans just jump on the bandwagon whenever the local team does well.

Baseball in my own heart and mind

My own affection for baseball came at a young age and through a combination of both rational and emotional attachments. I am able to recall the summer of 1957 when I knew baseball was always going to rank first in my heart when it came to sports in my life. Before then I had played pickup games in the park and knew some of the rules, and that summer I would join my first Little League team. I had spent the spring throwing the ball on our porch to my mother—bruising her leg several times and probably breaking a window or two. I had heard of my father's exploits in semi-pro leagues. I heard everyone laugh when my mother told the story of dropping me as a baby through the bleachers when my father hit a game-winning homerun. However, it was later that it would become obvious that baseball was not my strongest sport to play—hitting, throwing, or even catching a curveball is harder than it looks.

However, none of those elements was primary in making baseball my game to love even though they contributed. For me, baseball's primary excitement came to me through the newspaper every morning, where the data-filled box scores and statistics lit up my life. As the 1957 season progressed and being an 8-year-old-boy learning to read and do arithmetic, my mind was perfectly tuned to one thing and one thing only—baseball data and statistics in the local newspaper at my doorstep each morning. All summer, I was up early waiting for the paper to arrive and soon would be devouring each number in the box score of my favorite team and then one-by-one all the box scores I could find to analyze. The Sunday paper was special—it had the stats of every player in the majors. First, I would check out my own stat sheets and make sure the newspaper had everything correct. Then I would analyze the performance of my favorite players and teams, followed by more analysis of the leaders in all the categories—batting average, at bats, hits, runs, home runs, steals, and runs batted in (RBIs). I loved the RBI data because my favorite player, Hank Aaron, often led the league in that category. In 1957, he had 27 more RBIs than the runner up and led the league in home runs—that's right, a country boy in New York State who loved numbers became a Milwaukee Braves fan because he loved to count Hank Aaron's RBIs. Meanwhile, baseball had taught me percentages, metrics, and even models for the lineups. And I must add that the sports journalists taught me how to read as well. I became skilled at consuming the sports sections of newspapers.

Over the years, it was thousands of little moments (statistical, physical, and emotional) that built upon each other that kept exciting my feelings for the grand game of baseball. I remember my gloves—the first one was small and used, and the second was bigger and new. My father's glove was tattered, stiff, and too big, but it was so cool just to wear it to show to my friends. I eventually had a fancy new bat that was too big and too heavy, but I didn't care.

My first time seeing baseball on television was for my 11th birthday, and it was the All Star game in 1960. My cousin, a fan of the American League, and I, a supporter of the National League (NL), sat on the couch in my living room, ate popcorn, and took in Major League Baseball (MLB) on my family's first new television. Four players from the Milwaukee Braves were on the starting NL team, and I was completely infatuated with baseball by the time the game was over, and all the popcorn was consumed.

Of course, I collected baseball cards, and since I liked the Milwaukee Braves, I had as many cards for Braves players as I had for the other 15 major league teams. That's right; there were only 16 MLB teams in 1958. There had been 16 teams for the entire period of 1901 to 1961—just a little over half of the 30 teams we have now. The Braves have stayed with me, even though I have been to only a few of their away games (at the old Shea Stadium with my children) and no home games. My Braves connection received a boost when WTBS superstation in Atlanta hit the airwaves, which meant Atlanta Braves baseball was on the cable network nearly every single night. My wife became a fan, and we watched Dale Murphy do marvelous things, while Larvell "Sugar Bear" Blanks, Biff Pocoroba, Rowland Office, and the rest of the Braves made us laugh. My baseball data obsession received a boost with the publication of *USA Today* and its *Baseball Weekly* in the 1990s. Soon everywhere you looked (papers, magazine racks, TV, and internet), there were baseball data and statistics. I knew my wife was going to be a good sport about baseball because when we were first married, we would walk down the beach in Hawaii and stop to watch little league games even though we didn't know anyone on the team.

I love baseball because, in addition to teaching me mathematics and how to enjoy popcorn, it gives me plenty of data to crunch, statistics to compute and contemplate, models to build and analyze, and strategies to consider. There is a strong, nearly magical connection between mathematics (mostly as data, statistics, and models) and baseball.

Baseball: an amazing sport with plentiful data

The game of baseball involves two competing teams with some unique features: no clock, the defense has the ball, there is equality in opportunity, and at least some of the players appear to be pretty average in physique— their special physical talents are hidden in their amazing coordination, eyesight, and quickness. Historically, the cultural settings of baseball started nearly 200 years ago—before the American Civil War—and given its longevity, one could say baseball is a sport for any era as it built popularity right from the start of its invention. In the quiet times in America, fans appreciate baseball's excitement. In the complex and hectic times, fans find solace and relaxation with the sport's slower tempo. The major leagues didn't start until 1876 (NL) and in 1901 for the American League (AL), but those 144+ years have given us plenty of data and many generations of enjoyment. Today those two leagues together form MLB, even though starting in 1973 they play a slightly different game with a designated hitter (DH) in one league and not the other. And then in 2000, the two leagues started merging their game by having interleague games. In a weird accommodation which is also in effect for the World Series, the home team's league determines whether the DH is used. Watching the AL pitchers bat just a few times in a season is sometimes funny or dangerous.

The early history of baseball is fascinating, with many issues to settle on field size and shape, the rules and processes of play, and the makeup of teams and equipment. Suffice it to say, baseball as a sport was being played in some fashion by many Americans at the time of the US Civil War (including soldiers from both sides of the war). And shortly thereafter, there were professional teams being formed. The early years were tumultuous and disorganized, with rival leagues and players often upset over restricted player movement between teams within a league (the player-reviled reserve clause).

On the elements of baseball strategy and tactics, the early years were dominated by small ball—walks, singles, bunts, hit-and-run, and stolen bases. The lack of power from the "dead" ball led to lower slugging averages, and pitchers challenged hitters without the threat of the long ball leading to ERAs near 2.50. As an ultimate example of the dead-ball era, the

Chicago White Sox entire team hit three home runs for the season of 1908, yet they finished with a record of 88-64.[1] Before 1921, the same baseball was used during all or most of the game—until it started to unravel. The cost-conscious owners had fans throw back the balls when they were hit into the stands. The new rule in 1921 was that baseballs had to be replaced when they were dirty.[2] With a cleaner ball in play, players hit better with a ball that acted better when pitched, could be seen more clearly, and traveled farther when hit. The result was many more homeruns, especially off the bat of remarkable slugger Babe Ruth. After that, baseball was played a new way—long ball.

The sport's operations allow for extra players to substitute as a relief pitcher, pinch hitter, pinch runner, or new fielder. The teams have special categories of players—active, injured, 25-man or 40-man rosters, and minor leaguers. The sport allows for different strategies and tactics by the teams, especially since the geometry of the outfield (distances to and heights of fences) can be substantially different. Today, only a few teams play small ball: walk, bunt, steal, hit behind the runner—and most teams play long ball: swing hard for home runs and extra-base hits. Some teams seek social cohesion, and others don't mind a divided clubhouse. Just how important is teamwork in baseball? Coordination and collaboration on double plays seem to be necessary, but in baseball, the All Star teams would beat the championship team because most of the contribution to the team comes from the quality of individual performance and not as much from coordinated teamwork.

Baseball: a game with wonderful statistics

History holds a significant place in baseball, especially in its statistics. There are a few numbers that most all fans remember and even some non-fans recognize—20–game winners, batting averages of .400, batters with 3000 hits, pitchers with 300 wins. Some of the more illustrious numbers are 56, 60, 714, 755, 2130, and 2632 (consecutive games with a hit by Joe DiMaggio, Babe Ruth single-season home run record that lasted 33 years, Babe Ruth's career home run record, Hank Aaron's career home run record that surpassed Ruth's, Lou Gehrig's consecutive games played record that lasted 56 years, and Cal Ripken, Jr's consecutive games played record, respectively). And to our purpose in this book, there are plenty more numbers for the grist of the sabermetrics mill—over 200,000 games, over 2 million runs and strikeouts, 4 million hits and over 300,000 home runs and steals, over 100 different franchises and teams, almost 20,000 players, and mounds more of statistical data for all of us to collect, sort, compute with, contemplate, and enjoy.[3]

Sabermetrics are for fans, coaches, GMs, and team owners (fantasy and real) to do their jobs or embolden their lives with mathematical intervention.[4] Some want to know about ingenious metrics to understand if their team is using the most efficient method ("smart baseball") to win. Some only want to know the sabermetrics math to help generate real wins in the standings. And, dare I write, some just want to see good baseball, and hardly care about the mathematics that produces the quality of play and the wins in the standings.

The measures of performance enable the players, coaches, managers, GMs, and fans to follow the game in a rational, quantitative way. Proper merit-based performance data can provide the difference between looking good and being good, short-term success and long-term success, and winning by good fortune or good talent.

Players have different skills. Scouting systems often use the five tools for field players to rate players—speed, power, arm strength, hitting for average, and fielding. Fans have different preferences for how these tools affect the games. A person who loves pitching duels is probably not going to enjoy games with scores of 9-6 with 5 home runs. Other fans come to see the home runs and the big scores. The run environment of the teams separates fans and cities. Different game strategies create differences in data (e.g., runs per game, strikeout rate, home run rate, and stolen base rate) and different run environments. Thinking that many fans do not like to stay at the ballpark too long, MLB has wanted to speed up the games. How to do that without altering the game and its statistics too much is a highly controversial question. Baseball does not want to lose the drawing power of its endearing statistics. Most baseball fans love their stats. Many fans understand scorekeeping, fill out score cards at the games, and, like me, read newspapers and internet sports sites just for the stats.

There are data—items counted during the game (at bats, hits, runs, RBIs, doubles, triples, home runs, putouts, assists, errors, strikes, balls, walks, strikeouts, earned runs, wins, losses, quality starts). There are statistical measures or metrics that can be directly calculated from the data (e.g., batting average, slugging percentage, on-base percentage, and on-base average plus slugging average, earned run average, strikeouts per game, walks plus hits divided by innings pitched, fielding percentage, and the strikeouts to walks ratio). There are models that come from assumptions (Pythagorean wins, predicted wins, predicted number of home runs, strength of schedule, defense independent pitching statistics), there are predictive analytics that help to determine who to put in the lineup, when to pinch hit, when to go to the bullpen. Sabermetrics can provide player value measures and statistical measures of success (runs created, runs prevented, WAR—wins above replacement). Sabermetrics has improved the calculus for the measures, metrics, and models used before and during games.[5] This book will explain the how and why of those improvements.

Math provides baseball with its reasoning and measures. The measures define the value of the players. The reasoning defines the value of decisions. Sabermetrics gives these measures, reasoning, and values their intellectual foundation. Managers can use the sabermetric measures to build the lineup, decide on replacements, and design inning tactics and game strategy. GMs have a big role in player salary and opportunity, and sabermetrics gives insight for determining accurate player value. Owners bring the money, but the GMs are the ones spending it on the players, the manager, the coaches, the scouts, and the staff. Since it is the players that make the most money and have the most influence over the team's success, sabermetrics can help decide which players to recruit, draft, trade for, trade away, release, keep, and pay, and how much to offer and pay. Teams that take advantage of the analytical measures of player performance using Sabermetrics to field their team and thereby outsmart competitors are said to be using "moneyball" in reference to the Oakland A's success with such a system presented in the book *Moneyball*.[6] These days, teams have analytic staffs helping with that decision-making. Baseball's arbitration system and luxury tax (competitive balance tax) also help establish player value and annual salary growth.

Baseball provides math with a natural system to define performance contributions and productive teamwork. Contribution in baseball comes primarily from quality of play of the individuals on the team. Teamwork comes from the models of baseball strategies and tactics (e.g., optimal lineups, positions, substitution) that result in the success of the team. While these elements of success can be complex and in many sports the data are often challenging to collect, these factors are more direct and obvious in baseball. It's not that baseball is simple—it's not. Instead, the reason is that the most significant baseball data are easy to collect and the performance measures can be tested and verified over a large data set. The meaningful performance accomplishments can be counted or determined, and the lineup, position, and substitute models are structured enough to be understood. These attributes make baseball ideal for sabermetric math and modeling to help the teams compete. At the same time, math benefits greatly by using baseball as a popular, well-understood example of the utility of math, the value of data, and the impact of modeling.

Baseball: a business with huge egos and a desire for profits

The monetary support for baseball comes from the fans and ultimately goes to the players and the owners, who have been in constant struggle over their shares. Much of the money comes from gate receipts, parking fees, consumables, and broadcast rights. Some of the money makes that trip from fans to players and owners in a less direct manner. For instance, there is considerable money made through tax breaks, advertising revenue, and high-value memorabilia.

There have been two long-term issues for the baseball business—the reserve clause, which bound a player to his team for as long as the team wanted even after the expiration of a contract, and MLB's exemption from antitrust laws. These two legal issues were the battleground between the players and the owners for well over a century. One is mostly resolved with player free agency, but, amazingly, the antitrust issue is still with us.

The exemption to **antitrust** laws was granted to **MLB** as a result of a suit from a rival league in 1915 and the eventual decision handed down by the Supreme Court in 1922. This allows MLB to be a monopolistic enterprise. The decision was challenged in subsequent cases in 1953 and 1972 but was not overturned.[7] The antitrust exemption has allowed MLB to engage in practices that would likely have been unacceptable in other situations. This exception is controversial to this day.

The reserve clause had been in the player contract since 1879. The contract stated that the rights to the player were retained by the team upon the contract's expiration. As a result, players were not free to sign with another team. Once signed to a contract, players could be reassigned, traded, sold, or released by the team, but the player himself could not initiate any moves on his own. The owners held all the power. After decades of hassles, strikes, and legal cases, the shift in the power structure between players and owners occurred in December 1975 when an arbitrator's decision brought an end to the primary effects of the reserve clause. The players had made gains with the formation of a union in the 1960s, but the owners held that all-important reserve clause. The big financial issue was that the reserve clause inhibited the player's salary. If a player and team could not come to terms, the team could renew the player's contract for 1 year. If a player didn't like the contract being offered, his only option was to hold out without pay.

Things began to change when Marvin Miller became the executive director of the Players Association in 1966. The pace of change quickened, and the players' influence strengthened. In 1969, Curt Flood's lawsuit against baseball challenged both the exemption from antitrust laws and the reserve clause. He lost his case, but Flood and his case built solidarity in the players. At almost the same time in another case, the Supreme Court reaffirmed baseball's antitrust exemption. The 1972 player strike was over pensions— not the reserve clause. There continued many issues of cat and mouse between the players and the owners. However, through the years, the players became more and more organized and the owners more and more afraid that the reserve clause would inevitably be lost. As the owners were losing their advantage, they refused to modify the reserve clause in collective bargaining, trying to hold on to the absolute power they had through the clause. Meanwhile, the players won more bargaining power through the negotiated process of salary arbitration. This didn't allow players to choose the club to play for, but it gave players higher salaries.

For several years in the 1970s, the owners were doing enough to avert a test case of the reserve clause, although the Catfish Hunter case became a big issue for both sides and the fans. Because the Oakland A's had not made payments to Hunter as scheduled during the 1974 season, Hunter, a 25-game winner and the Cy Young Award winner, became the most valuable free agent ever. The result for Hunter was a 5-year, $3 million contract with the Yankees. The result for all the other players was seeing how much more money free agency could be worth to them.

There were two reasonable cases to test the reserve clause in 1975.[8] The Dodgers had renewed Andy Messersmith, who held out for a no-trade clause. And Dave McNally, who had retired after beginning the season with the Expos, was put on a renewed contract. If McNally wanted to play again, he could claim free agency after 1 year as stated in the player contract. The Players Association filed grievances after the 1975 season, and the hearings took place. The heart of the issue of the reserve clause came down to the mathematical definition of the number "one" that appeared in everyone's contract. To the players, "one" meant the number 1—a single year (only one time). To the owners, the "one" in the contract gave them the right to iterate—a rolling number of 1-year renewals forever. The three-person arbitration committee ruled for the players, and free agency was born. Not only were Messersmith and McNally free agents, but also any player who had a renewed contract or held out for a year would be free to negotiate with other teams.

The new collective bargaining agreement defined the workings of free agency but not until the owners had locked out spring training. Cooler heads prevailed, and the 1976 regular season started on schedule as negotiations continued. An agreement was reached with free agency for players with 6 years in the majors after playing 1 last year under a renewed contract along with compensation for teams losing free agents. Players' salaries went up dramatically. The next strike took out some of the 1981 season, resulting in free agents being able to negotiate with all the teams. In 1994 an attempt by the owners to impose a salary cap resulted in a strike that ended the season and wiped out the playoffs and World Series and delayed the beginning of the 1995 season. The result was that owners were able to create a payroll tax system that taxed teams with salaries above a limit, but more importantly, many fans thought the strike was wrong, given the high salaries of the players and the riches of the owners. Attendance dipped for a bit, and some fans left and never came back to the sport. However, the sport and game won many fans back. The business of baseball seemed healthy with most everyone making money as the fans' appetite and monetary support for the game keep rising and rising. However, the rough times of the pandemic of 2020 seems to have brought out the old player-owner wounds over how to spread the wealth.

Baseball as entertainment

People love attending the games in person. Baseball resonates with American fans—all ages and all the people. However, because of admission costs and a long distance to a stadium, most people watch the games on television or listen on radio. In-stadium attendance peaked in 2012 at nearly 80 million, and even though attendance dipped down to 68 million in 2019, over 110 million fans watched games on television and digital streaming. MLB has its own streaming system.[9] Most significantly, over 170 million people call themselves baseball fans. And many kids got interested in baseball through on-line, video games, or fantasy leagues, even when kids' interest in playing Little League baseball was declining.

A factor in the decline of stadium attendees is that the average time for a game has ballooned to over 3 hours. Some of the extra time is caused by slow play, but another contributor is that more relief pitchers are being brought into games, and that procedure produces a pause in play. People want to see their players in action, not a manager talking and a pitcher warming up. Fans love their favorite star players doing well and watching their pursuit of records and milestones. *Sabermetrics* and math help fans track that progress. A chase for a home run record takes the prize for being the biggest draw of fans to come to the park and tune into television, the computer, or their cell phone.

Baseball cards used to be the biggest deal in many kids' lives—that interest is not so intense today. Jersey and memorabilia sales have taken over people's interests with a more significant revenue stream. Baseball card companies now combine their card sales with memorabilia—uniform jerseys, bats, gloves, and caps. As players become popular, their jersey becomes a best seller. All Stars, most valuable players, home run hitters, strikeout leaders, and most any statistical leader attract fans' interest in jerseys and more money for the league, the teams, and the players association where all the players share equally in the sales revenue. Today's fans bask in a more stylish form of fantasy by wearing baseball-connected designer clothes. Baseball cards were never fashion statements.

Recreational fun

Baseball is an activity that can be fun in several ways. In my own young life, we had recreational fun in pickup games with whoever was at the park. If we had bases, a backstop, and a fence great. If not, no problem; we still played. 3 on 3 was fine, 5 on 4 was better, 6 on 8 was OK. We never had 9 on 9 for a pickup game. If it was 1 on 1, we could play a form of Home Run Derby or just take turns counting the bases you could accumulate in an inning. The object of recreational baseball is fun, and I am sure we always had that. We never worried much about winning or losing nor whether we were following the precise rules of baseball. If any rule made sense and seemed to increase fun for the kids playing the game, then it became part of baseball for the duration of that game but was long forgotten by the next group of pickup game players. Sometimes the score was kept, but it was never very precise. And the score was forgotten by the time the next pickup game started. We all pretended to be our favorite player—even though, in my small, MLB-remote village, most kids had never seen or watched a major league game in person or on the television. That is vastly different today. Kids and adults know the players, have seen them play or appear in a commercial, and, likely, go to the games wearing a symbol of that player or his team.

Let's go back to the rules for a moment. Pickup games back in the 1950s were crazy, chaotic events. There was no telling what could happen when a ball was pitched or hit—or when a player was running or throwing. Balls went to strange places, and two runners were just as likely to be on the same base as different ones. That was one of the reasons it was fun. It was also why some games took longer than they should have. Even without umpires, there always seemed to be a lot to discuss, debate about, maybe get a little angry, and definitely laugh about after nearly every play. Personalities were, are, and will be developed, earned, and refined on the baseball fields of America.

Organized leagues like Little League, Pony League, American Legion, Babe Ruth League, and Cal Ripken Baseball were fun too, except sometimes adults spoiled some of the fun with too many rules and too much competition. That adult interference probably happens today even more frequently. Fortunately, some of baseball's resurgence stems from a recent increase in Little League participation, which has been rising since 2013.[10] Girls were allowed to play on Little League baseball teams starting in 1974. There are also programs to develop fields and teams for inner-city youth. MLB sponsors its own program named **Reviving Baseball in Inner Cities (RBI)**. In addition to youth baseball, there are several national adult and semi-pro leagues where some recreation and more gritty action takes place for adults who like to recreate and compete on ball diamonds.

Baseball as a developer of teamwork and fitness in youth

Baseball has always attracted youth to join teams to learn, practice, and play together to develop fitness, coordination, team skills, and confidence. As was mentioned earlier, special physical attributes are not necessary to play the game, so many children have an opportunity to play baseball in some form in their communities or on school teams. Baseball skills can be developed in back yards, porches (with some danger to parents' legs and windows), and parks. You just need interested children, a baseball, and gloves. Playing catch and fielding grounders and pop flies are fun and developmental. As the young players join more structured and organized teams, they see the benefits of hard work in practice, support for and from their teammates and community, sportsmanship, and the benefit from the cognitive elements of the game.

Baseball as a major cultural influencer

Fans often identify with players because the baseball uniform and equipment do not block or hinder their identity. Fans see the players on television as people and relate with them. From that connection comes cultural influences that enhance appreciation for both homophily and diversity. Players can and do become role models of successful people of different or same races and ethnicities. Unfortunately, there can be a negative cultural element of that connection as well—the contamination of the role-model influence when players are identified as taking performance-enhancing drugs (PEDs), cheating, exhibiting racism, bad sportsmanship, and flaunting their salary or their privilege.

Over the decades of baseball, thousands of players have been in favorable and detrimental situations when serving as positive or negative role models. The popularity of the sport puts the players in those situations, even if the players themselves do not seek out those roles. Lou Gehrig and Cal Ripken are considered hardworking survivors for their long games-played streaks. Babe Ruth and Mickey Mantle are considered talented, carefree, eccentric characters for their outstanding play and excessive partying lifestyles. Dale Murphy and Roberto Clemente are considered kind, caring players of high moral character, while Ty Cobb and Alex Rodriguez are at the other side of that measure of good character. Mark "The Bird" Fidrych, Bill "Spaceman" Lee, and Willie "The Hotdog" Montanez added spice and character to the game. Jackie Robinson and Hank Greenberg were considered strong-willed and high-character pioneers for their roles in breaking racial and religious barriers in baseball. There were more baseball people than just the players who became cultural influencers—announcers Bob Uecker and Harry Caray (and many others) boosted the popularity of the game and humanized the sport and its people, and talented columnists and book writers such as George Plimpton, John Kieran, and Frank Deford (some of my favorites) gave baseball and sports their deeper context.

I can't let my mention of George Plimpton go by without adding Plimpton's role in the most amazing, all-time best, April Fools' joke.[11] He wrote a fictional article entitled "The Curious Case of Sidd Finch" that appeared in *Sports Illustrated* on April 1st, 1985. In the article, Plimpton described Sidd as a Mets rookie who could throw 168 mile-per-hour strikes. The hoax caused the baseball world to go crazy trying to find, interview, and report on this amazing (nonexistent) player. Of course, many aftereffects have taken place since that day, all in the good fun of baseball's nuttiness and Plimpton's sense of humor.

However, for many decades there was a dark cloud over baseball and the US as time after time black players were denied the opportunity to play in the major leagues. Player and Manager Cap Anson and Commissioner Kennesaw Mountain Landis were significant influencers and decision-makers, who despite many denials to the contrary, prevented the integration of baseball. The owners never codified their discrimination but instead had a secretive "gentlemen's agreement" to maintain the color barrier. Racism was widespread beyond the owners as players, coaches, and fans also played roles to segregate the game. This cultural stain kept black players in separate leagues until 1947 when Jackie Robinson, recruited by General Manager Branch Rickey, joined the Dodgers. Up until then, the most lucrative and successful effort for black players was the Negro League run by Rube Walker from 1920 to the late 1940s. Even after 1947, it took another 12 years for every team to integrate and finally give black players a fair opportunity to play MLB. Just think of the quality of play that was missed by fans, like me in 1957, over the decades of an unjust "gentleman's agreement."

America's national pastime

Why is baseball so popular in the United States? As was mentioned, the sport was played early in our history and played significant roles at various times in US history. It caught on as a fun way for local businesses and industries to compete against each other in local leagues. It doesn't take many resources except a flat open field to set up a diamond infield and as much outfield as you wanted to allow. The United States has plenty of space for fields of green grass. None of the expenses for baseball are major outlays for communities—a backstop and a fence help the play but are not mandatory. And the players need gloves, and the team needs a few bats. The big expendables are baseballs, and for many years, even in the major leagues, balls were used until they fell apart. In modern times, that is not the case. A major league game uses 100 to 120 balls on average.[12]

The sport just seemed to fit the American psyche—in the open air and refreshing. Parents taking their children to the games was a rite of passage in America. Its popularity increased year after year, and by 1900, baseball was the American pastime—the most popular game to play, the most popular sport to watch, the most popular subject to talk about. When you added hotdogs and bottled beer to baseball, Americans were hooked. The major leagues had the right mix of drama, controversy, heroes, and cultural phenomena. Fans loved or became enthralled with Joe DiMaggio hitting in 56 consecutive games, Jackie Robinson and his agreement with Branch Rickey not to strike back at his attackers, the Black Sox scandal, the greatness of Babe Ruth and Satchel Paige, the hustle of Ty Cobb and Pete Rose, the power of Mark McGwire and Sammy Sosa, the records of Hank Aaron and Barry Bonds, the rivalries of Yankees and Red Sox—Cubs and Cardinals—Dodgers and Giants (east and west coast versions), the energy of Casey Stengel and Earl Weaver, and even the drama of Billy Martin and George Steinbrenner ("The Boss" but not Bruce Springsteen). However, the pastime was not in its full prosperity until the inclusion of black Americans in the late 1940s and 1950s as the color barriers were finally breached and American baseball included the best players in the world. Just a look at the players' pictures in one of my favorite books, *Baseball's Greatest*, shows that black players have had a tremendous impact on the quality of play.[13]

Then there are the main events that annually attract interest from people around the world beyond the regular fans. There are celebrations on opening day with pageantry and a renewed hope for a successful season as "hope springs eternal." The annual All Star game with its home run hitting contest shows off the best players in the league. Then the playoffs and World Series step by step create a championship team for the season. Baseball keeps special data for these events that produce their own records (e.g., many fans know the New York Yankees have won 27 World Series, and Yankee catcher Yogi Berra played in 10 of those). These are the high-quality grist for the national pastime mill.

I think it is true that the Baseball Hall of Fame (HoF), located in Cooperstown, NY, and opened in 1939, is the most celebrated of its kind of sports shrine. And, in several ways, it could be the most controversial. Members of the HoF are elected by the Baseball Writers Association of America or a special veterans committee. Inclusion on 75% of the ballots is the criterion for selection. The statistical metrics of the candidates are paramount—you don't make the Hall of Fame without excellent statistics. And over time, *Sabermetrics* provides the most compelling evidence and measures of performance. In many ways, the HoF has become as much the Hall of Stats as the Hall of Fame. There are over 330 members as of 2020. The controversy isn't so much about the statistics but the behavior of the candidates. Sometimes there are complications from inappropriate behavior—if the misbehavior is related to baseball that can be a nonqualifier (e.g., betting on games) and if not (e.g., crimes against society), the player may still make the vote. Perhaps an example of misbehavior being ignored is Ty Cobb's inclusion in the HoF on its very first ballot in 1936 (3 years before the building opened) with the highest vote percentage for the first 56 years of selections. The recent debate is which group (qualified or nonqualified) do players linked to steroids belong. Should these PED-enhanced players be in the HoF or not? Time, *Sabermetrics*, and the collective judgment of the writers and veterans on the committees will determine that answer.

The future of baseball

The game, sport, business, entertainment, recreation, cultural influence, and pastime of baseball will continue to grow and change. Today's issues concern the length of games, should there be a pitch clock, the validity of the shift, relief pitchers starting games as openers, how to prevent stealing of signs, and should robotic machines call balls and strikes. The future will bring many new issues, just as perplexing. However, there will be some things that do not change. Baseball will still be America's pastime—played by many, watched by many more, and appreciated by nearly all Americans. *Sabermetrics* will improve as modeling and analytics use artificial intelligence and machine learning to improve predictions and assessments of player value. GMs, managers, owners (real and fantasy), players, and fans will have more insight into the game. And despite that progress, since humans will still play the game, wildcard teams will sometimes win the World Series, sometimes teams will go from worst to first and vice versa, once in a while unheard of rookies will win batting and pitching titles, and even the Mariners (in Seattle or not) will eventually win a pennant and the World Series. And unsurprisingly, whatever happens in baseball will be talked about on television, living rooms, computers, chat rooms, water coolers, and corner bars, and will be written about in newspapers, on the internet, and in books, and the mementos of those future events will appear in the museum collocated with the Hall of Fame at Cooperstown. And most importantly, parents will take their children to the game to eat peanuts and popcorn and learn to fill out scorecards, and kids will have fun playing ball in porches, parks, fields, and computer screens. And if one of those kids is named Sidd Finch, watch out for his fastball.

Problems to consider

1. For many years the MLB consisted of 16 teams and now there are 30. Does this increase in number of teams and players dilute the talent in MLB? What are other factors that affect talent level? Design a measure for the level of talent in MLB over the years. According to your model, what era was the talent at its highest level?

2. What years of a player's performance should be considered when establishing the value of a player in a fantasy league? Explain why and how you determined this result.

3. Do you think baseball is using the correct dimensions for its infield? What factors could you use to make your decision?

4. Many people consider Joe DiMaggio's 56 game hit streak established in 1941 as the most unbreakable record in baseball. What are the factors that make a baseball record a challenge to break?

5. There is talk today about baseball's run environment—how many **runs score** on average in a game. There is speculation that fans in general prefer more runs. What rules can be modified to increase the run environment? Can baseball increase run production without increasing the length of the game or ruining the legacy of its numerical records?

6. Some pitchers save their best pitch during an at bat for a punch-out third strike. Others want the hitter to swing poorly at a pitch early in the count and weakly hit the ball that is likely an out, and therefore use their best pitch early in the count. How can you determine a pitcher's best pitch?

References

1. https://www.baseball-reference.com/leagues/AL/1908.shtml
2. Schwarz, Alan. "History of Rule Changes." *ESPN*, February 4, 2020. Available at: http://a.espncdn.com/mlb/columns/schwarz_alan/1503763.html
3. https://www.baseball-reference.com/leagues/
4. Albert, Jim. "Sabermetrics: The Past, the Present, and the Future." In Joseph Gallian, editor, *Mathematics and Sports*. Mathematical Association of America, 2010.
5. Costa, Gabriel, Michael Huber, and John Saccoman. *Reasoning with Sabermetrics*. McFarland and Company, 2012.
6. Lewis, Michel. Moneyball: The Art of Winning an Unfair Game. W. W. Norton, 2004.
7. https://www.theantitrustattorney.com/baseball-and-the-antitrust-laws part-iii-baseball-reaches-the-supreme-court/

8. Thornley, Stew. "The Demise of the Reserve Clause: The Players' Path to Freedom." Available at: https://milkeespress.com/reserveclause.html

9. Adler, David. "MLB Sees Fan Growth Across the Board in 2019." *MLB.com*, September 20, 2019. Available at: https://www.mlb.com/news/mlb-increased-viewership-attendance-in-2019

10. https://www.littleleague.org/who-we-are/history/

11. Erkstine, Chris. "Column: Remembering Sidd Finch, the Mets Prospect Who Seemed Almost Too Good to Be True." Los Angeles Times, April 1, 2020. Available at: https://www.latimes.com/sports/story/2020-04-01/sidd-finch-mets-prospect-george-plimpton-baseball

12. https://www.foxsports.com/other/story/major-league-baseballs-have-a-short-shelf-life-062912

13. Syken, Bill, editor. *Baseball's Greatest*. New York: Sports Illustrated Books, 2003.

Is baseball still the national pastime?

2

Chapter outline

Introduction

Baseball has gone through many changes over the years. Baseball was our National Pastime; however, it had rocky years over the decades, too. This chapter will deal with the popularity of baseball over the last century (1920–2019). We will compare its strengths with the likes of the National Football League and the National Basketball Association, the two other favorite team sports in America. So let us take a look at America love affair with our National Pastime.

"Baseball is played everywhere: in parks and playgrounds, prison yards, in back alleys and farmers' fields; by small boys, and old men, raw amateurs and millionaire professionals. It is a leisurely game that demands blinding speed; the only game in which the defense has the ball. It follows the seasons, beginning each year with the fond expectancy of springtime and ending with the hard facts of autumn."[1]

"Baseball was first coined the National Pastime in the mid-19th century, when the first professional league was organized in 1871. It's called a pastime for a reason. Unlike football and basketball, baseball has been around since the 19th century. Although professional baseball had yet to be established, the sport was around during the Civil War. "Being the first professional sport is strong grounds for being called our nation's pastime."[2]

Sabermetrics. DOI: 10.1016/B978-0-12-822345-1.00002-7

So, is it still the national pastime today? This chapter will deal with that question. We will interject how sports have changed over the decades. This chapter will deal with the emergence of other sports and their effects on baseball, television and its impact on the game, the pace of the game, wagering on sports, marketing of the game and its star players, player salaries, loyalty to teams, and parity.

Hello, my name is John DeSomma and I grew up in Hoboken, NJ, as a lover of baseball. My dad was a huge Yankee fan and Joe DiMaggio (Joltin' Joe) was his hero. My dad took me to my first game at the "old" Yankee Stadium (The House That Ruth Built) as an eight-year-old. Everyone who loves the game of baseball has experienced a time or a place where they fell in love with the sport. Now think back to when you fell in love with the game! Here is my story. As a seven-year-old boy in 1956, I first heard of a Yankee star centerfielder and the fact that he won the Triple Crown. I remember my dad talking about the 1956 World Series, Don Larsen's perfect game, and a great catch by a player named Mickey Mantle to save the day. I decided that my desire was to see Mickey play in person.

It was a beautiful Sunday afternoon, June 23, 1957, when my dad took me to my first baseball game at: The "House That Ruth Built." The Yanks were playing the Chicago White Sox in a doubleheader. I was happy to be with my dad, and my first chance to see Mickey play in person. In the top of the first inning, with the Sox already ahead by a run, Larry Doby hit a ball to deep right–center, and Mickey tracked it down and made a leaping catch by the auxiliary scoreboard. In the second game, with the Yanks trailing 4-0 in the bottom of the ninth, my dad and I moved to box seats right behind home plate. There were two runners on base when Mickey came to bat. He was larger than life to me with those muscular arms, broad shoulders, and the huge number 7 on his back. In that at bat, Mickey hit a 3–run homer off Dick Donovan two-thirds of the way into the upper–deck in right field. I could not believe that an individual could hit a ball that far, especially in the old Yankee Stadium. That day, thanks to my dad, I became a lover of the Yankees, Mickey Mantle, and baseball.

America's love affair starts

Elysian Field in Hoboken, New Jersey, is believed to be the site of the first organized baseball game, giving Hoboken a strong claim to be the birthplace of organized baseball. The game was played on June 19, 1846. The teams that played were the New York Nine versus New York Knickerbockers. The New York Knickerbockers were rumored to have played games in 1845; however, the 1846 game is long recognized as the first officially recorded baseball game in US history.

In a real "squeaker" the New York Nine defeated the Knickerbockers 23-1 in four innings. My guess is that this first game was a prophetic look at how a future game between the 1927 Yankee team and the 1962 Met team would turn out. As a "Hobokenite" knowing the first organized game was played in my hometown makes me feel even better about the game that I love.

Americans have always loved baseball. From the beginning, baseball was beloved because it was a game that, unlike other sports, you could play no matter your size (or strength for that matter). Baseball is a game of skill and strategy, like chess, and it didn't cost a lot of money to participate or attend a game at the ballpark to watch the best play the game. It's a unique game where there is no clock and the only major sport where the defense controls the ball and thus the game. A game where in the vast majority of cases you would not be injured. It is a game children, teenagers, and adults gravitate to quickly. Baseball is truly an American game. It has survived two World Wars, a gambling scandal that would have destroyed other sports, and a depression.

Let's look back at baseball through the decades of the 20th century, starting with the 1920s, to see if and when in the last 100 years baseball's popularity declined as the national pastime.

Here are several examples of what we will discuss in this chapter:
- baseball's milestones achieved through the decades;
- television and its impact on the game;
- the pace of the game;
- emergence of other major sports;
- performance-enhancing drugs (PEDs);
- marketing of the game and its star players;
- player salaries;
- loyalty;
- parity.

The 1920s

What made baseball great in the 1920s:
- It had a super, superstar (George Herman "Babe" Ruth).
- Baseball had a dominant team (Yankees, six pennants in the decade).
- Gameday prices were affordable.
- The radio brought baseball into your living room.

Baseball was no doubt the national pastime in the 1920s. The most famous athlete in the United States at that time was baseball star Babe Ruth, the right fielder for the New York Yankees. The colorful Ruth hit more home runs than any player had ever hit before. He excited fans with his outgoing personality.

The most popular sports in the 1920s were baseball, boxing, collegiate basketball, and football, but other sports also attracted vast interest such as ice hockey and tennis. For the first time, large numbers of Americans began to pay money to watch other people compete in athletic contests. More people went to baseball games, more people followed baseball, and more people played baseball for fun than any other sport during this period. Ruth was the perfect hero for the Roaring Twenties. Also, the 1920s was a decade when college football became more popular, too.

I am starting with the roaring 1920s as baseball was the king of sports at that time. In the 1920s, baseball had a cornucopia of stars. The 1920s gave us some of the best players of all time: Babe Ruth, Rogers Hornsby, Tris Speaker, and Pie Traynor to name a few. However, baseball commenced the 1920s with the Black Sox scandal. The public believed their players should be honest and have integrity. The scandal happened because miserly owner Charles Comiskey was not true to his word to his players and a notorious gambler, Arnold Rothstein, took advantage of the situation. The great team of the 1919 White Sox was the lowest paid team in the league. The Black Sox scandal was a Major League Baseball game–fixing scandal in which eight members of the Chicago White Sox were accused of throwing the 1919 World Series against the Cincinnati Reds in exchange for money from a gambling syndicate led by Rothstein. As a result of the scandal, baseball was in serious trouble. The owners appointed Judge Kenesaw Mountain Landis as the first Commissioner of baseball, with absolute control over the sport to restore its integrity. Landis had a reputation as a tough disciplinarian. Here are two famous quotes from Judge Landis after he took over as commissioner:

"Baseball is something more than a game to an American boy; it is his training field for life's work. Destroy his faith in its squareness and honesty and you have destroyed something more; you have planted suspicion of all things in his heart."[3]

"If a jury of your peers finds you not guilty, I will reinstate you back into baseball."[4] Obviously, Landis was so upset with the way the trial was handled that he went back on his word. Despite the White Sox player acquittals in a public trial in 1921, Judge Landis permanently banned from professional baseball all eight men involved. The punishment was eventually defined by the Baseball Hall of Fame to include banishment from consideration for the Hall of Fame. Despite requests for reinstatement in the decades that followed—particularly in the case of Shoeless Joe Jackson, one of the best players of his time and who batted .375 in the series and played flawless defensively—the ban remained in place.[5]

Just as the sport was about to be splintered, a man named George Herman "Babe" Ruth came along. Babe was a young man who was raised in a Baltimore orphanage where he developed an extraordinary ability to throw and hit a baseball. After being traded from the Boston Red Sox to the New York Yankees for cash in 1919, he revolutionized the game by introducing frequent and massive home runs, the likes of which were not seen before. Ruth outhomered teams. The fans loved him and what he was accomplishing. Because of Ruth's accomplishments, the scandal was soon forgotten and baseball once again became the focus of sports in America.

The 1920s had its share of headlines in this decade. They included:

- Babe Ruth played his first game for the New York Yankees in 1920 after being sold by the Boston Red Sox.
- A year later in 1921 baseball is heard on the radio for the first time, piquing fan interest.
- Also in 1921, Ruth hits an astonishing 59 homers. Astonishing because his total home runs was more than all the other teams in the American League. It has been argued that 1921 was Ruth's best year as a hitter.
- In 1927, the Babe breaks his record by hitting a record 60 homers. This record would stand for 34 years.
- Finally, in 1929, the Babe hits his 500th home run. At that time, no individual was even close to that achievement. After the 1930 season, the Babe had more than twice the number of home runs as any other major leaguer. By the way, the nearest player was Rogers Hornsby at 275.

It is obvious that the Babe dominated the headlines in the 1920s. Baseball attendance was up 47% in the 1920s and the national pastime title was intact! However, with the collapse of the stock market in October 1929, baseball had a new challenge in the upcoming decade. The pace of the game was good, too. By the way, the average time of the game in the 1920s was just under two hours.

The 1930s

In the midst of the Depression of the 1930s, all sports in our country were affected. Understandably, paid attendance at the ballparks was down. However, star players of the 1930s including Bill Dickey, Lou Gehrig, Charlie Gehringer, Joe DiMaggio and pitchers like Lefty Grove, Carl Hubbell, Lefty Gomez, and Dizzy Dean kept piquing the fan's interest. In addition, Babe Ruth calling his shot in the 1932 fall classic against the Cubs was a major headline and is still being debated today, almost 90 years later. Other major accomplishments of the decade were:

- The first All Star game was played in 1933.
- The nickname "Gashouse Gang" was applied to the Cardinals team of 1934.
- In 1935 the first night game was played at Crosley Field in Cincinnati, Ohio.

- In 1936 the Baseball Hall of Fame was opened in Cooperstown, New York.
- Johnny Vander Meer pitched back–to–back no hitters in 1938.

Let's take a look at these events in detail that kept baseball at the top of sports in the 1930s.

The first All Star game was played on July 6, 1933, at Comiskey Park in Chicago, Illinois. The starting and winning pitcher for the AL was Lefty Gomez. The catcher was Rick Ferrell, Lou Gehrig was at first base, Charlie Gehringer at second, Jimmy Dykes at third, shortstop was Joe Cronin, and the outfield was Ben Chapman, Al Simmons, and Babe Ruth. Not a bad team would you say?

The NL starters were starter and losing pitcher Bill Hallahan, catcher Jimmie Wilson, Bill Terry at first base, Frankie Frisch at second, Pepper Martin at third, and to round out the infield, Dick Bartell as shortstop. The outfield was Chick Hafey, Wally Berger, and Chuck Klein.

The AL won the game 4-2, with Babe Ruth being the star (what a surprise). The Babe hit a 2-run homer in the third inning and made a great catch at the wall in the eighth inning. A footnote to the game: both teams' managers (John McGraw and Connie Mack), five out of six coaches, and two out of the four umpires on the field that day would be future Hall of Famers.[6]

The following year in 1934, the St. Louis Cardinals had a scrappy team and the city of St. Louis and the country were mesmerized by their antics. It was common knowledge throughout the National League that the Cardinals often went out onto the field with dirty, rancid smelling uniform, adding to their "scrappy" image. Appropriately, their irascible shortstop, Leo Durocher, dubbed the team "The Gashouse Gang".

The 1934 Cardinals were an excellent team that went on to win the NL pennant by winning 18 of their last 23 games. Then they defeated the Detroit Tigers in seven games with Dizzy Dean beating submarine pitcher Elden Auker 11-0 in Game 7.

Another milestone of baseball in the 1930s was the first night game in major league history. This game took place on May 24, 1935, at Crosley Field in Cincinnati when the Cincinnati Reds beat the Philadelphia Phillies 2-1. This game has affected baseball to the present day. While there were few games started at night that year, baseball today is primarily a night-televised sport. Millions more can now watch a professional game after work, keeping baseball in the public eye every day. In the 1930s this was a progressive move for baseball to take. To finish off the decade, the great Yankee teams of the late 1930s went on to win four straight championships.

Johnny Vander Meer's back–to–back no–hitters in June of 1938 were milestones that captivated the country. On June 11 this 23-year-old pitcher beat the Boston Bees at Crosley Field. He walked only three batters that day.

Later that week at Brooklyn's Ebbets Field, Johnny no–hit the Dodgers. He was wild that day, walking eight. Vander Meer almost blew the game in the ninth inning when he walked the bases loaded. He also needed a few spectacular defensive plays by his teammates to ensure his no–hitter. No pitcher has duplicated that feat. It may never be equaled; never mind pitching three no–hitters in a row. It is a record that will probably never be broken. Unfortunately, Johnny never regained that mastery again. He did finish his career with 3.44 ERA, which is good in any decade.

Professional baseball in the 1930s helped to keep the nation's mind off the horrible Depression, at least for a few hours a day. Finally, the Yankees great teams of the 1930s ended the decade with four straight World Series, which was the most consecutive titles in baseball history up to that point.

Attendance dropped 8% as compared to the beginning of the decade. But considering the Depression, which lasted into the 1940s, that drop was not of big significance. By the end of the 1930s, baseball, which started off so poorly with the Depression, was still considered our national pastime because of the players and events that took place. However, football was coming along strongly with college football leading the way. The National Basketball Association had not yet been established.

The 1940s

At the start of the 1940s, baseball kept right on rolling along. The 1940s saw baseball hit new heights. Ted Williams, "Teddy Ballgame," hit an amazing .406 in 1941. Amazing because no player since 1941 has matched that feat. Ted Williams' .406 has not been topped in nearly 80 years. The closest to .400 was Tony Gwynn at .394 in 1994's strike-shortened season and .390 by George Brett in 1980. Two honorable mentions of .388 go to Ted Williams in 1957 and Rod Carew in 1977. Also in 1941, Joe DiMaggio's 56–game hitting streak was established and is still intact. DiMaggio's streak was followed all over the country. The daily question of baseball followers in 1941 was, "Did Joe get a hit today?" These two feats propelled baseball into the 1940s and in large part kept baseball on top.

With America entering World War II all sports took a blow. However, at the insistence of President Roosevelt, professional baseball continued for the duration of the war. However, with the best players doing their patriotic service, the lowly St. Louis Browns won their only pennant in 1944. When baseball returned with their star players the game peaked again.

In addition, the early 1940s brought professional women's baseball to the masses when the first formal women's professional baseball league was established. The All-American Girls Professional Baseball League (AAGPBL) first took the field in 1943. The league was founded by Phillip

K. Wrigley. Yes, he was the chewing gum king and the owner of the Chicago Cubs of the National League. The All-American Girls Professional Baseball League lasted a dozen years and gave more than 600 women an opportunity that had never existed before. Baseball was innovative and the American people enjoyed it.

In 1947, baseball hit its biggest home run with the signing of Jackie Roosevelt Robinson to a professional contract with the Brooklyn Dodgers. This signing broke a 78-year-old travesty when Branch Rickey, who was a player, a manager, and GM signed Jackie Robinson to a contract. Rickey is also credited with the establishment of the farm system when he was the GM for the Cardinals. However, when he took the GM job for the Dodgers, he orchestrated the biggest moment in baseball history.

Baseball, the sport that I loved, was a purely white sport at the professional level. I didn't notice because by the time I started to love baseball in the mid- to late-1950s minorities were in the game. However, when I think back on the years before 1947 and realized what had happened it saddens me and it should sadden you. I can't imagine what Jackie went through in his first years in professional baseball. Jackie was called names every day of his life. Death threats to him and his family. Yelling at him, constantly, and calling him all sorts of despicable names, and telling him he was not human that he was an ape. Players going out of their way to injure him or pick a fight. These men were cowards because they knew Jackie couldn't fight back. Some of these men are in the Hall of Fame, including the first commissioner of baseball, Judge Landis. It was a travesty to vote "no" on racial issues because of the color of one's skin. To me that was reprehensible. I believe segregation in baseball was the most grievous of all problems baseball had ever encountered.

After Judge Landis—who ruled baseball with an iron fist and propitiated the gentlemen's agreement among owners (no minorities in baseball)—died, the game had a chance to do something great. After Landis' death, Albert Benjamin "Happy" Chandler, a politician from Kentucky, took over the reins. The time was right for a momentous change.

On April 15, 1947, which was ranked as the number 1 moment in baseball history, Jackie Robinson played Major League Baseball. Cheating by steroids, gambling, or using technological advancement to spy on your opponents are the closest other degrading acts to segregation. Cheating brings you to a lowly state and possible embarrassment and would hurt the sport. But doing despicable things to another human being is the most reprehensible. Because of ballplayers' and the owners' racist attitude, the vast majority of the public missed out on seeing the likes of Josh Gibson, Cool Papa Bell, and Buck O'Neil. At least for the next generation, we didn't miss out on ballplayers like Willie Mays, Roberto Clemente, Frank Robinson, Hank Aaron and many others just because of racist men.

The delay in bringing minorities into the major leagues was finally over. Baseball was back on the rise with minority players adding great excitement to the game. As the 1940s drew to a close, baseball was still extremely popular and still the national pastime. A new, major sport was entering the professional arena as the National Basketball Association was formed in 1946. In 1948, a poll asking about the popularity of sports indicated that baseball was still number 1, followed by football and basketball. As we moved into the 1950s baseball was still the king of sports.

The 1950s

The decade commenced with the "Whiz Kids" of Philadelphia winning the pennant in 1950, followed by Bobby Thomson's 3–run homer to capture the NL pennant in 1951. The Thomson homer was in the bottom of the ninth of the final two out of three playoff games to decide who would go to the World Series to play the hated Yankees. Thomson's homer off Ralph Branca was later called "the shot heard around the world." It capped off a brilliant 3–game playoff win and one of the greatest comebacks in NL play.

In mid-August the Giants were trailing the Dodgers by 13½ games. They went on a winning streak and caught the Dodgers on the last day of the season. Who hasn't heard Russ Hodges' maniacal call: "The Giants win the pennant! The Giants win the pennant! The Giants win the pennant! The Giants win the pennant!" Another great moment for baseball.

In addition, the 1950s were a decade of movement. No franchise had changed cities since the Baltimore Orioles had become the New York Highlanders in 1903 (and changed their name to the Yankees a decade later). However, in 1955 the St. Louis Browns moved to Baltimore, but the major franchise moves of the Giants and Dodgers in 1958 to California ranks in the top 20 moments that shaped MLB's history. The Brooklyn Dodgers and the New York Giants moved from New York to Los Angeles and San Francisco, respectively. While it devastated the National League rooters in New York, the moves opened up baseball coast to coast, and we can all see baseball achievements because of these relocations.

The 1950s was also a decade of firsts starting with the Brooklyn Dodgers winning their first and only World Series in 1955 against the New York Yankees. In 1956, these two teams met again and Don Larsen did what no pitcher before or since accomplished. On October 8, 1956, he pitched a perfect game in the World Series.

New York City and its surrounding communities are a huge market for sports, which helped baseball stay at the top of all sports. In one 10-year period (1949 to 1958), a New York team was in every World Series and had superstars. That certainly didn't hurt when a survey was taken in early 1960 and baseball was Number 1, again.

However, there was trouble on the horizon. On December 27, 1958, there was an NFL championship game between the Baltimore Colts, piloted by Johnny Unitas, and the New York Giants, quarterbacked by Charlie Connolly. Because the game was televised nationally and it was the first championship game to go into overtime, the audience was huge. When Alan "The Horse" Ameche scored on a third down and goal from the 1-yard run with 6:45 left in the overtime period, the Colts won 23-17. This Championship game would later be called the greatest game ever played. It catapulted the NFL to new heights and started to erode baseball's popularity.

The 1960s

The 1960s was a turbulent time in our country and in baseball. The 1960s was all about expansion. Baseball would go north of the border to Montreal and west again to San Diego where the Padres were born. Earlier that decade, the New York Metropolitans (Mets) and the Houston Colt .45s clubs were established.

Houston changed their name to the Astros in 1965 when their new stadium was finished. The "eighth wonder of the world," the Astrodome was the first indoor stadium for sports. Another progressive move to attract fans and stir interest in the game.

In the 1960s, baseball had its share of excitement. The decade started with another Top 20 moment in baseball history when Pittsburgh Pirate Bill Mazeroski hit a Yankees' Ralph Terry pitch over the left–field fence for a game winning homer in the 1960 World Series, the first World Series to end on a walk–off homer. Excitement swept fans off their feet in 1961 when the M&M boys (Mantle and Maris) battled for the single–season home run record. They hit 115 homers between them, breaking the record of 107 set by Ruth and Gehrig in 1927. Roger Maris did surpass Ruth with 61 homers, however, it was done in 162 games. The commissioner at the time, Ford Frick, ruled that in order for the homer record to be broken it must be done in 154 games. The ruling stayed that way until 1991 when an eight–member panel, established by Commissioner Fay Vincent, ruled that Maris' total stood as the record.

The asterisk was dead even though there was never an asterisk, just a parenthetical statement which read "in 162–game schedule." Roger Maris passed away in December 1985 never knowing his record would stand. Looking back, another sad moment in baseball history.

In 1961, the American League expanded to a 162–game schedule with the addition of two teams, the Washington Senators, who later that decade became the Minnesota Twins, and the Los Angles Angels owned by Gene Autry.

There were 20 teams now, after the National League expansion in 1962, but only one in each league would make the postseason. This system of play was not good for baseball and contributed to a laissez–faire attitude toward the game. From 1960 to 1964 only one team in the American League made it to the World Series. That team was the New York Yankees.

In 1965, the amateur draft was conceived which would allow the top high school and college players to be drafted by MLB. The draft was driven by ownership to block the Yankees, who had the money, from buying top ballplayers from other clubs. The draft and the fact that the Yankees had an aging team accomplished the objective. In fact, in two years, the Yankees finished in 10th place, dead last in the American League. In the beginning of the 1960s baseball's vision had deteriorated, and you know what happens if your vision is not focused; without a progressive vision you would dwell carelessly. Baseball's vision needed to be lifted to a new level in order to stay competitive with other sports.

One of the significant changes in the game of baseball came in 1966 when Marvin Miller became the executive director of the Major League Baseball Players Association, a position he served in from 1966 to 1982. Under Miller's direction, the player's union was transformed into one of the strongest unions in the United States. The biggest accomplishment during Miller's tenure was striking down the reserve clause and creating free agency. Former MLB Commissioner Fay Vincent said upon learning of Miller's death in 2012, "I think he's the most important baseball figure of the last 50 years."[7]

In 1969, Curt Flood challenged the reserve clause after a trade to the Philadelphia Phillies. The reserve clause was the core of all baseball contracts. The reserve clause was part of all players' contracts and stated the rights to players were retained by the team until the contract's expiration. Players, under these contracts, were not free to enter into another contract with another team. Every player signed the same type of contract. Flood made his case against the reserve clause. The union reps voted to support Flood, but Miller warned Flood that they both were making powerful enemies for life. Miller also warned Flood that his career would be over and that he would likely never be voted into the Baseball Hall of Fame. As of 2020, that last statement remains true. My personal feelings on Curt are that he should be in the Hall of Fame when you combine his statistics as a player and his courage to sacrifice his career to fight the system.

However, it wasn't until 1975 after a couple unsuccessful attempts that the courts ruled the reserve clause was unfair to the players. Free agency was born, with Jim "Catfish" Hunter being the first free agent and signing with the Yankees. This was a great win for the players, but perhaps not a great move for baseball in the long run as we shall discuss later.

In a 1965 poll of the favorite American sport, football surpassed baseball for the first time as the most popular sport. In the sports background, during the first half of the 1960s football's popularity was growing with the introduction of the American Football League (AFL) in 1960. Teams were established coast to coast and the pace of the game was fast and loose. Their offensive stars included Joe "Willie" Namath, Len Dawson, and Daryl Lamonica.

With football rising in popularity in the 1960s and a merger of the two leagues in place, the AFL and the NFL agreed to have a championship game between the two leagues; this game would later be called the Super Bowl. This event, played in the dead of winter, turned into the most watched sporting event of the year. The build-up and the hype of playing one game for all the marbles was a stroke of marketing genius. The only games equal to that would be a Game Seven of the World Series or a game 7 NBA final.

For the first time, in the middle of this decade, football surpassed baseball in popularity according to a Harris Poll. I think a main reason for this had to do with the number of games played in a season. In football, nearly every game has some sort of significant impact. A team could get into the playoffs with a .500 record. Baseball purists would have been terribly disappointed if that happened in baseball. Pete Rozelle's parity narrative was working and the quote "on any given Sunday" was born. The fans loved that their home team, even with a mediocre record, could still make the playoffs and win it all.

Under the merger agreement of the AFL and NFL, announced on June 8, 1966, the new league would be called the NFL and split into two conferences: the American Football Conference (AFC) and the National Football Conference (NFC). All eight of the original AFL teams would be absorbed by the NFL. However, the actual merger didn't take place until before the start of the 1970 season.

Adding to the popularity of football in the 1960s was a team that was not very good in the late 1950s but turned into a powerhouse in the 1960s under the tutelage of Vince Lombardi. The Green Bay Packers from 1961 to 1967 won five championships. This small Wisconsin town, Green Bay, became the mecca of football in the 1960s.

Basketball had its day in the sun in the 1960s with a 100–point individual game effort by Wilt Chamberlain in March of 1962. However, the game was played in the small town of Hershey, Pennsylvania. Only 4100 fans attended the game in the small arena and there is no film footage of the game. Wilt accomplished 6 of the top 10 scoring games in league history, each time with at least 70 points.

The Celtics owned the 1960s. They won 9 of 10 championships in the 1960s, winning 7 in a row. The only year they were not champions was 1967. In addition, basketball was gaining momentum in popularity too.

Basketball popularity was driven by the play of the Celtics' Bill Russell, and the Celtics were, by far, the dominant team of the NBA in the 1960s. The battle for the most popular in the 1960s sports world was intensifying.

The 1970s

At the beginning of this decade football was now entrenched as the number 1 most popular major sport interest in America. Football had a quick pace and star power. In the early 1970s baseball was floundering and didn't expand its fan base. Marketing was not high on their list of priorities in the early 1970s either.

Meanwhile, in football the playoff system was expanded to two teams in 1972. Every Sunday for a vast majority of the season a win by a team was crucial. This piqued fan interest in cities all across America.

The Super Bowl, as it was now called, drew even more and more interest. Rivalries were developed and teams rose to the top, such as the Pittsburgh Steelers, Miami Dolphins, and the Dallas Cowboys. The NFL marketed the game very well. The Super Bowl was the most watched championship game of all sports. One game for all the marbles.

Baseball went through its paces in the 1970s, too. There were great moments and seasons: Pete Rose's 44 – game hitting streak, the Big Red Machine of the mid – 1970s, the designated hitter rule in 1973, and Reggie Jackson's 3 – homer night in the 1977 World Series from which he got his nickname "Mr. October." However, the biggest moment, which captivated the country, happened at the beginning of the 1974 season. On April 8, 1974, Henry Louis Aaron (Hank Aaron) hit an Al Downing pitch over the left-field wall for his 715th homer, breaking the record of the immortal George Herman Ruth.

Aesthetically, baseball went through some of its most radical changes during the 1970s. The trend of multipurpose stadiums made artificial turf more of a necessity than a novelty. By decade's end, nearly half of the teams were playing on the "carpet." The new stadiums, all in the NL, were also known as the "cookie cutter" ballparks. They had symmetrical dimensions and Astroturf, thereby losing the ambiance of the old parks.

Baseball in the 1970s produced four dominant teams: the Pirates, champions in 1970 and 1979; the Oakland Athletics from 1972 through 1974; the Reds in 1975 and 1976; and, the Yankees of the 1977 and 1978 vintage. The only other team to win a World Championship that decade was the Baltimore Orioles. However, 1972 was a sad year in baseball as Jackie Robinson and Roberto Clemente passed away.

In 1973, the designated hitter was instituted, a bold move by the American League. Also in 1973, George Steinbrenner led investors to purchase the

Yankees for less than $10 million. Talk about a vision, George had it. He transformed the most successful club that had recently slumped under Columbia Broadcasting System (CBS) ownership, to a pennant-winning and then a championship club in less than four years. The Yankee brass have orchestrated deals and used the Yankee name to boost the franchise, which is now worth an estimated $4.6 billion. Only the Dallas Cowboys are worth more at $5 billion. However, when you add in the YES network's net worth, which the Yankees reacquired in March 2019 for $3.47 billion, you have a total net worth of over $8 billion. Yes, that was a progressive vision and the end goal has still not been achieved.

Despite these moments in the 1970s baseball still trailed football in popularity when this decade ended. By 1976 baseball was no longer the only form of athletic entertainment for Americans. The NFL merged with the AFL in 1967, followed by the NBA–ABA merger in 1976. Basically, the 1970s Major Leagues were marching in place while the NFL and NBA took giant leaps, marketing their superstar players.

The 1980s

Now we jump to the 1980s where the Major Leagues had some catching up to do. However, the 1980s was a decade of transition for baseball. Power numbers were noticeably down, making the eye–opening numbers of the steroid-enhanced decades coming up look even gaudier. In the 1980s, players hit 40 home runs just 13 times, the lowest of any decade since Babe Ruth revolutionized the game. The fans wanted to see the long ball and what they got was John McGraw's early 1920s version of the game.

When you examine the facts, the 1980s was not a good decade for baseball. This is not what baseball needed. Three work stoppages were recorded, with the 1981 season losing a total of 712 games. The 1981 season was juggled and baseball for the first time had a split season, with a first–half winner playing a second–half winner in both leagues. The fans were not happy as the rhythm of the season was gone. In a long 162–game baseball season this factor is important.

With a power shortage in the 1980s the fans wanted to see the long ball again. However, in 1985, they had a chance to see baseball history. On September 11, 1985, Pete Rose hit an Eric Show pitch into left center for a single, passing the immortal Ty Cobb as the all–time hits leader at 4192. Baseball rejoiced for Pete Rose. (Charlie Hustle made it to the top of the hits mountain.)

Two other events happened in the 1980s that didn't do baseball any favors. They both happened in 1989. The first one was Pete Rose's banishment from baseball because he gambled on baseball while managing the Cincinnati

Reds. The Dowd Report, initiated by Commissioner Bart Giamatti, made it clear that Rose bet on games, even games he managed. When confronted with the evidence, Rose refused to meet with Giamatti or refute the charges. The Commissioner had no choice but to give the ultimate punishment in baseball. What a shame and disgrace only four years earlier, Rose was celebrated as the king of hits, now he was banished as a corruptible lying gambler.

Also that year, on October 17th an earthquake struck Candlestick Park in San Francisco before Game Three of the 1989 World Series. Although there were many deaths and $5 billion in property damage, no one at the ballpark was hurt. However, the World Series was postponed 10 days to restore power, and more importantly it allowed engineers to check for structural damage to Candlestick Park. The A's went on to win both Games Three and Four to sweep the series.

At this time steroids started to rear its ugly head. Football and basketball had their share of problems in the 1980s but baseball took a punch that landed on the chin. After the mediocre decade of the 1970s, the 1980s was not what the doctor ordered.

The 1990s (the steroid era begins to bloom)

The steroid era brought rise to controversy and discouragement among fans. By the late 1990s Mark McGuire and Sammy Sosa were powering their way through the record books, setting the stage for Barry Bonds to blow their records away. This negativity surrounding baseball combined with the NBA's resurgence caused the NBA to pass Major League Baseball in terms of popularity. Baseball in the 1990s was now number three in popularity behind football and basketball.

The 1990s in baseball did have some highlights. In 1995 Cal Ripken surpassed Lou Gehrig's record of 2131 consecutive games played. Baseball fans loved it. In 1997 baseball honored Jackie Robinson for his triumph over segregation in baseball. On the 50th anniversary of his first game played, Jackie's number 42 was retired from all major and minor league teams. Baseball again moved forward by honoring the past. Baseball has been the best of all the sports in doing that very thing.

Interleague play was finally introduced to Major League Baseball during this decade (starting in 1997) as well as an expanded playoff system, going from two playoff teams in each league to four. This playoff format was included in other major sports for years but not baseball. Again, a bold step for baseball. Unfortunately, baseball usually lagged behind the other major sports with innovative ideas and was hesitant with ideas that could backfire.

Can anyone say androstenedione or PEDs, known by the acronym PEDs? Androstenedione is a steroidal hormone that enhances strength by raising male hormones to build muscle, to speed recovery times, and to build strength rapidly. By the mid 1990s, androstenedione was in every clubhouse in the majors.

This era, commencing with the 1990s, was the darkest in years for baseball, probably since the Black Sox scandal 100 years ago. All of a sudden fans started hearing the word *androstenedione* (andro for short), a substance that was found in just about every clubhouse in the MLB in the early 1990s. The decade closed on what was thought of as a high note. In 1998, both Mark McGuire and Sammy Sosa passed Roger Maris' single–season home run record. They hit 70 and 66 homers, respectively. Andro fueled the debate of whether these records were legitimate or tainted by PEDs. Remember, andro was not a banned substance in the 1990s.

You may ask, why would baseball players jeopardize their careers and possibly their health by using PEDs? Simple—for fame and financial gain! After the strike in 1994, which crippled baseball, sluggers especially thought it was worth the risk. Besides, the players who were putting up these fantastic numbers were getting the big bucks in their next contract.

Baseball went through a terrible year in 1994. The baseball player's strike in August ended the season. Because of the impasse between owners and players, the World Series was canceled. It was the first time in 90 years that a World Series wasn't played. Fans were upset and baseball had to make it right. Why wasn't the possible PED problem handled right away? My theory, just my opinion, is that as far back as the mid-1990s baseball's brass knew what was going on with the players. However, with the gaudy offensive numbers being put up in baseball, the owners and the commissioner's office turned a blind eye to the problem. Baseball took a huge hit in 1994 and management wanted to get the fans to return. The home run was the way. PEDs and a possible juiced ball pioneered the way. Fans went back to the ballpark and baseball's popularity was on the rise again.

The 21st century

The new millennium started with Barry Bonds ramping up his efforts and Major League Baseball marched into the 21st century. When Mark McGwire broke Roger Maris' home run record, we were too enamored with the long ball and in too much denial to admit McGwire had cheated. Once baseball came to its senses and shunned McGwire, we had a new knight in shining armor who was going to take the record back from a cheater.

Barry Bonds broke McGwire's record just three years later, saving us from having a liar hold the all–time home run record. Oh, how we were naive. When everyone realized that Bonds' head looked like a small watermelon, we were again angry that someone who used PEDs not only had the single–season home run record, but the all–time home run record as well. If Bonds was not using PEDs then the only other plausible explanation was he must have been lifting weights with his ears.

In addition, MLB was the last of the four major North American professional sports leagues to implement an instant replay review system. (What a surprise, last!) Instant replay review was first implemented during the 2008 season. Under that system, only the umpire crew chief could initiate a review, and one or more members of the umpiring crew would review the video at the stadium and render the decision to uphold or overturn the call. Only boundary home run calls could be reviewed, either if the initial call was a home run but might not have been or if the initial call was not a home run but might have been.

Technological advances changed the landscape of MLB. The current instant replay system was implemented in the 2014 season. Under the current system, each manager is allotted one challenge per game, with additional challenges granted only if the previous one was successful. From the eighth inning on, the umpire crew chief is allowed to initiate his own replay review. The umpire crew chief is also allowed to initiate a review during any inning if the play in question is a boundary home run call. This system, in the vast majority of the cases, would get the call right, however, the challenges could add as much as 10 minutes to an already long game. The replay system needs to be tweaked to allow faster decisions on the field while still getting the call right.

Just think if instant replay was part of the game in the last 35 years. Don Denkinger's blown 1985 World Series call that turned the tide for the Royals, Richie Garcia's missed call in a 1996 playoff game which helped start the Yankees to another dynasty, and finally Jim Joyce's terrible call denying Armando Galarraga a perfect game. Baseball was not going to overturn that call. Joyce admitted he had blown the call and with Armando getting the next batter out, in my opinion Robert Manfred missed a chance for baseball to restore credibility with the fans. Only 21 perfect games were thrown in baseball in the common era. Armando technically got 28 up and 28 down and still did not get a perfect game.

Football began making inroads to become the most popular sport as far back as the 1970s, and two decades later, it wasn't much of a contest anymore. If baseball still had the hearts and minds of the American sports fan, then football could lay claim to everything else.

At the start of the new millennium, football was by far the most popular sport. However, Americans still called baseball the national pastime. Why? Do they do it out of habit or because baseball is still the most loved sport? By any measure—TV viewership, merchandise sold, fan surveys—football is more popular. Football may be more popular, but it doesn't capture the heart as baseball does.

Football may not be superior to baseball, but it's undeniably more popular. In January 2020, Harris Poll Interactive reported that 34% of Americans identify professional football as their favorite sport compared to only 16% for baseball. When one adds in the 11% that say college football is their favorite, it's clear that Americans prefer their pigskin game over the ball with the stitches. It is not just viewership that is on the rise for football. These days, kids are choosing "more exciting" sports over baseball. Just a few short years ago, the number of kids aged 7 to 17 playing baseball fell 24%. This is especially true with city kids.

When I was growing up, on a beautiful summer's morning you couldn't wait to get a bat and a ball and play just about all day. Despite growing concerns about the long–term effects of concussions, participation in youth tackle football has soared 21% over the last decade. The Sporting Goods Manufacturing Association, an industry trade group, said baseball participation fell 12.7% for the overall population. Wikipedia, the wizards of football, had another marketing trick. As the sport increased in popularity, they expanded game coverage. Monday night football, Sunday night games, and, now, Thursday night football on cable all across the country. This has kept football in the limelight during a time when most sports lovers are home. All I'm saying is for the baseball enthusiasts to examine the game and see where it can be improved without desecrating this great game. The NFL and NBA did that very well. Marketing of any sport in the 21st century must be global. Baseball as of 2020 is behind. The NFL and NBA started serious marketing globally many years ago. Just this year the Cardinals will host a 2020 home game in Mexico City.

Baseball needs innovative thinking in the next decade, but changes will not come that quickly. Let's take the NBA as an example. The NBA has been around for about 75 years; we often forget how long it took for basketball to become a staple in the American and global sports scene. The NBA was established in 1946 and it didn't reach its peak until the 1990s. Some of its rule changes were a gamble but turned out for the best. The biggest was the introduction of the 24-second clock in the 1954 to 1955 season.

Following the NFL merger, the NBA and the newly formed ABA merged in 1976. This brought about the three–point shot which the fans loved and it transformed the ending of an NBA game.

Baseball needs to address two issues immediately. First, the designated hitter rule. You cannot have different sets of rules in the game; either abolish it or make it mandatory for both leagues. My opinion would be to install it in both leagues. I'm jaded on this issue because when I first started to watch baseball, pitchers for the most part couldn't come close to the Mendoza line (0.200), In fact, my whole argument can be wrapped up in a single pitcher of the 1960s. His name was Hank Aguirre. Good old Hank, the poster boy for the sacrifice bunt. Old Hank did muster 33 hits in his career! Unfortunately, that was over 16 years. His lifetime batting average was a whopping 0.085. He would need a catapult just to get to the Mendoza line. On the positive side, when he was at bat one could get a "cold one" without missing any action. In 1908 when Jack Norworth wrote "Take Me Out to the Ballgame," he was actually thinking about Hank's prowess at the plate. You know, "one, two, three strikes you're out at the old ballgame." A side note to having two sets of rules on batting is the fact that it gives the NL an advantage in inter-league play. They have batted the entire season. It's certainly an advantage even if it is a small one. Seriously, having the pitcher bat is going to bring fans back? Finally, name me one other sport that has two sets of rules? This is not cool!

The other situation that needs addressing immediately is the pace of the game. Major League Baseball has to do something to speed up its pace of play or it will continue to lose fans. Some games drag on with pitchers taking forever to get set, having frequent mound visits, pitching changes, and batters constantly stepping out of the box. The days of Mike Hargrove should be long over and forgotten.

With the advancement of technology, most aspects of life are quicker than a couple decades ago. We want things done instantly. How many times have you gotten upset when your computer takes more than one second to respond? Well, that's what the fans want today. Why would people want to sit through a 3- to 4-hour baseball game when they could be watching 2 hours to 3 hours of an up–and–down basketball game?

In the NBA, you know you're going to get a shot every 24 seconds; in football a play every 25 to 40 seconds. In baseball, you may get two pitches in that time. That's a big difference. The time of a nine–inning game reached a record length in the major leagues in the 2019 season. The final figure for the 2019 regular season was 3 hours, 5 minutes, 35 seconds. That topped the 3:05:11 in 2017. Compare that to just under 2 hours in baseball's early rise to the top. Yes, that's a 50% increase.

Take a look at the World Series ratings from roughly the past 20 years and you will see a decline in fans watching the fall classic. People just aren't as interested in baseball as they used to be. Are steroids the cause? I believe it was one main contributor. The other major one was PEDs.

For the past three years (2017 to 2019) the NBA Finals have had higher ratings than the World Series, cementing the fact that more people are watching the NBA than MLB. So why are less people watching MLB than the NBA? Glad you asked. The game is just more exciting.

Baseball at times is tedious to watch. Even being at a ballpark could be a tedious task. Following the game at the park is hard at times unless you have prime seats. Especially true if you are bringing a young child with you. I would ask, how many times have you as a child or adult jumped up on a fly ball and it only was a short fly to center? A solution could be to have the home team's radio announcer broadcast his or her broadcast over the ballpark's public address system. This action could keep younger fans interested in the game itself.

Bryce Harper and Mike Trout have set up baseball for a great future on the field. However, there are so many more young superstars in the NBA than in baseball. These NBA guys are both young and superstars, the players in baseball are just young with potential to become superstars. Also, do you see baseball players promoting the sport on television? Are they doing commercials? No, that needs to change.

The pace of the game must be dealt with immediately. Baseball's ideas to speed up the game are moving in the right direction with the pitch clock and a relief pitcher having to face at least three batters. However, other ideas need to be implemented. Very important: No more work stoppages. If there is a work stoppage after the 2021 season, baseball may not recover. It's amazing to the average working man that millionaire owners and yes ballplayers can't reach an agreement of salaries and overall dollars shared. Actually, there have been eight stoppages in baseball since 1972. In five of the stoppages no regular season games were lost. However, in 1981, the owners and players differences caused 712 games to be missed. In the 1994 strike, baseball lost more than games, they lost fans and the ones that stayed true to the game were less enthusiastic. It enhanced the chances to explore PEDs, which the players did in droves.

During this century two major curses have been broken. The curse of the Bambino and the Billy Goat. In 2004 the Red Sox won their first World Series since 1918. The Sox came back from a 3-0 deficit in the and defeated the Yankees four games to three, then swept the Cardinals in the World Series. Ironically, the Sox have won the most World Series this century with four. In 2016, the Cubs won their first World Series in 108 years by coming back from a 3-1 deficit against the Cleveland Indians. Two great comebacks.

For me, baseball will always be my national pastime. I truly love the game. Yes, football wrestled the title of most popular sport title long ago. My dear friend and I talked several years ago about the "footballization of baseball." My buddy was right. Change is key when advancing a sport.

I am sorry to all the purists out there, but baseball needs to change with the times otherwise it is doomed to stay a second–tier sport.

Baseball may not be Number 1 anymore in popularity but with over 40 million people watching Game Seven of the 2019 World Series it isn't fading away either. It's still a great sport and with some changes it may be back on top.

References

1. www.wikiquote.org/wki/baseball-documentary
2. www.breezejmu.org/sports/double-take-is-baseball-still
3. quote fancy.com>quote>Kennesaw-mountain-Landis Wikipedia.
4. quotefancy.com/quote//quote/1597447/Kennesaw-Mountain-Landis
5. Wikipedia.org/wiki/black sox scandal.
6. Wikipedia library.answerthepublic.net
7. Washington Post/sports/2019/12/08

Baseball before steroids

3

Chapter Outline

When the steroid era in baseball began is a matter of debate, though most people agree that it was probably late in the 1980s. In 1991 Major League Baseball (MLB) placed steroids on the banned list but did not begin testing until 2003. In 2005 stricter penalties began to be imposed, and the era "ended."

Why was the ban on steroids not enforced for such a long time? There are many contributing factors. The Players Association opposed testing and MLB was afraid of a player strike. The Players Association was happy with the increase in average salary paid to the players, while the owners were happy with higher attendance and more revenue from TV contracts. Why did attendance increase? According to a saying attributed to Ralph Kiner, "Singles hitters drive Fords, home run hitters drive Cadillacs." Fans love a home run, and home run production increased dramatically during the steroid era.

But even before the steroid era, baseball was "America's Pastime." "Baseball" is played in many different forms by children, from pick-up games played in sandlots, to stickball played on city streets, to organized leagues such as Little League. As we grew older, we learned to root, root, root for the home team. Baseball purists would argue which player is better—among current players is Mike Trout better than Mookie Betts? But the real enthusiasts want to compare players from different eras—is Mike Trout better than Babe Ruth? Steroid use has tainted such comparisons.

Sabermetrics. DOI: http://dx.doi.org/10.1016/B978-0-12-822345-1.00003-9

While the basic rules of baseball have not changed much through the years, there have been many changes to the conditions under which MLB was played. The period from 1901 to 1919 was considered the dead ball era; before 1935 all games were played during the day; in the early 1940s many players served in the military; in 1947 the color barrier was broken; in 1958 baseball moved west (before that teams traveled mainly by train); in 1961 the American League (AL) expanded from 8 to 10 teams and increased the number of games from 154 to 162—the National League (NL) followed suit the next year. Further expansions followed, and more teams moved west, south, and north (to Canada).

Why is there not the same problem with comparing players from these different eras as there is with players from the steroid era? While conditions may have been different, all players for instance in the dead ball era played under equal circumstances, except for of course differences in natural talent. You could compare players from different eras by looking at how they performed with respect to the average player of their era. On the other hand, only some players used steroids which gave them a big advantage over the non-users. Their stats, especially power stats, were inflated and no longer could be considered viable.

It must be mentioned that performance enhancing drugs had been a part of baseball since the 1940s, namely amphetamines. It is thought that players returning to the game after World War II introduced them into the game. Baseball had the toughest schedule in professional sports: the number of games played, playing night after night, playing day games after night games, and doubleheaders. It is not a surprise that players needed something to counter fatigue. One difference between amphetamines and steroids was that amphetamines were taken with the intent of restoring an athlete's skills while steroids enhance the athlete's skills. It was thought that most players used amphetamines.

In this work we look at the history of baseball, its players, and their records through the decades from 1901 though the end of the 1980s. As the majority of steroid users are hitters, we will concentrate on offensive production, particularly home run production. We will only consider offensive production for players after 1901.

The NL was formed in 1876, the AL in 1901. MLB was founded in 1903; the two leagues remained separate but cooperated. At that time there were eight teams in each league. The first World Series was played in 1903 with the Boston Americans (later Red Sox) of the AL defeating the Pittsburgh Pirates of the NL five games to three (in 1903 it was a best of nine series). There was no series in 1904, but thereafter the World Series was played every year except 1994, when it was cancelled due to a player strike.

The Dead Ball Era

The period from 1901 to 1919 was considered the dead ball era. The balls used during this time were expensive to make. The same ball was usually used for up to 100 or more pitches, until they began to unravel. Balls that were hit into the stands, fair or foul, were thrown back. As a result, the balls, which were soft to begin with, got even softer the more they were used, making it more difficult to hit for distance. Home runs were few and far between.

Harry Davis was the definition of a power hitter during the decade 1900 to 1909, leading the AL in home runs four consecutive years from 1904 to 1907 with the prolific totals of 12, 8, 10, and 8 respectively. Clifford Cravath led the NL in home runs six times during the 8-year stretch when he hit his career total of 116, including setting the record with 24 for the Phillies in 1915, taking advantage of the small dimensions of his home park Baker's Field, where he hit 19 home runs that year. Frank "Home Run" Baker earned the nickname by leading the AL in home runs four consecutive years, with a high of 12 in 1913. In 1918, a starting pitcher and part-time outfielder for the Boston Red Sox tied for the league lead in home runs with 11. The next year Babe Ruth was converted to a full-time outfielder and set the record for home runs in a single season with 29, more than doubling his career total. His popularity ushered in the end of the dead ball era.

While the power hitters mentioned above, with the exception of Babe Ruth, are not well known, several all-time hitters played during this period. Ty Cobb had a lifetime batting average of .367, the highest of all time; he hit .420 in 1911. He is second all-time with 4191 hits, fourth with 723 doubles, and second with 297 triples. Tris Speaker had a lifetime batting average of .345, good for fifth all-time. He also hit 792 doubles, which is first by almost 50. Honus Wagner was known for more than the most expensive trading card. He was a lifetime .329 hitter and had 3430 hits, seventh all-time.

It should not be surprising that some of the greatest pitchers of all time played during this era. Walter Johnson won 417 games, second most all-time. He had a career ERA of 2.17, and completed 531 of the 666 games he started, with 110 of them ending in shutouts. Christy Mathewson won 373 games, had an ERA of 2.13, and completed 434 of the 551 games he started, with 79 shutouts. Grover Cleveland Alexander also won 373 games, had an ERA of 2.56, and completed 437 of the 600 games he started. Let us not forget Cy Young, the all-time leader in losses with 316. Of course, he could lose that many games because he won 511 games, 94 more than Johnson. A portion of his wins and losses came before 1901. In 1956 the award for the best pitcher was named in his honor. These are the four winningest pitchers in major league history.

The 1920s

Before the 1920 season Babe Ruth went from the Boston Red Sox to the New York Yankees, traded for No No Nanette—not an untouchable pitcher but a Broadway musical. As the saying goes, the rest is history. The 1920s were the decade of Ruth, a decade in which he led MLB in home runs eight times. In 1920 he shattered the home run record with 54. To show the dominance of this figure, the next highest total was 19. But even more extraordinary was that his total surpassed that of every other team in the AL and all but the Philadelphia Phillies, who hit a total of 64, in the NL. Ruth would once again break the single season record with 59 in 1921 and then 60 in 1927, a record that would stand until finally broken by Roger Maris with 61 in '61. Ruth's 1927 total was higher than every other team in the AL with the exception of the rest of his Yankee teammates. Lou Gehrig had 47 of the 98 hit by Yankee players other than Ruth. Gehrig, who replaced Wally Pipp at first base in June of 1925 and remained there for 3120 consecutive games, and Ruth were the foundations of the Yankees lineup, known as Murderer's Row. In 1927 Gehrig set the record for runs batted in with 175, breaking the record of 171 that Ruth had set in 1921. Gehrig's total is even more noteworthy when you take into account that he came up with the bases empty at least 60 times. During the decade Ruth hit a total of 467 home runs, which if that were his career total, would place him 36th on the all-time list. He hit 40 or more home runs eight times, that total was only achieved five other times during the decade. By the end of the decade Ruth had amassed more than 500 home runs; he was the only player to have more than 300 home runs. Ruth was more than a power hitter; his batting average during the 1920s was .355.

The second leading home run hitter during the 1920s was Rogers Hornsby. While hitting 250 home runs in the decade, leading MLB the two times Ruth did not, his yearly totals were inconsistent. He led with 42 home runs in 1922, then hit a total of 42 over the next two seasons, and then again led all of MLB with 39 home runs in 1925. Hornsby is best known for winning the NL batting title seven times during the decade, three times hitting over 400. He hit .382 during this period and .358 for his career, second best of all time.

The 1930s

The 1930s brought new names to the top of the home run lists. Ruth's domination of the home run totals had begun to diminish as his career was winding down. He managed to tie his teammate Lou Gehrig for the major league lead with 46 home runs in 1931 and was second the other 3 years from 1930 to

1933. By the time he retired on May 30, 1935 his home run total was 714, more than doubling Gehrig, who had hit 353.

Hack Wilson hit 56 home runs in 1930 and set the all-time record for RBIs with 191. Jimmie Foxx led the AL in home runs four times during the 1930s, hitting 50 plus home runs twice, 58 in 1932 and 50 in 1938, though his 50 home runs fell well short of the 58 home runs hit by Hank Greenberg. Foxx hit 415 home runs, batted .336, and averaged 138 RBIs during the decade. Lou Gehrig was one of the most consistent hitters in baseball history. He led the AL in home runs three times during the decade, hitting his career high of 49 in both 1934 and 1936. His total of 347 was second highest during the '30s. When he retired prematurely in 1938 due to the disease that bears his name, his career total of 493 home runs was second behind Ruth. Six times from 1930 to 1937, Gehrig had over 150 RBIs, including the AL record of 184 in 1931. He hit 4 home runs in a game in 1931 and hit 16 of his 23 grand slams during the decade. During the 1930s Mel Ott led the NL in home runs five times, though he never reached a total of 40. He hit 308 of his career total of 511.

During the decade, the 50-home run total was reached 4 times, while 40 home runs were reached another 14 times.

Two of the greatest hitters of all time began their careers at the end of the decade. Joe DiMaggio joined the Yankees in 1936 and led the major leagues in home runs with 46 in 1937. Ted Williams joined the Red Sox in 1939 and became the first rookie to lead the AL in RBIs with 145.

There were two other important events that occurred during the '30s. The first All-Star game was played on July 6, 1933 at Comiskey Park in Chicago, with the AL winning 4 to 2, behind a home run by Babe Ruth. It was supposed to be a one-time event, as part of Chicago's Century of Progress Exhibition. The game proved so popular that it has been played every year since. In 1936 the first five players—Ty Cobb, Babe Ruth, Honus Wagner, Christy Mathewson, and Walter Johnson (in order of vote totals)— were voted into the Hall of Fame.

The 1940s

The leading home run hitters from 1940 to 1949 were Ted Williams with 234, Johnny Mize with 217, and Bill Nicholson with 211. The low numbers of the leaders in home runs during the 1940s is attributed to World War II. Both Ted Williams and Johnny Mize lost 3 years of baseball in their prime due to military service. Ted Williams led the AL in home runs four times, twice before his military service and twice upon his return. In 1941, besides leading the league with 37 home runs, Williams batted .406, the last major league player to hit .400. Remarkably he was runner up to

Joe DiMaggio—who hit in 56 straight games—in the MVP voting. Bill Nicholson played all 10 years and led the NL in home runs in 1943 and 1944. Ralph Kiner led the NL with 23 in 1946, his rookie season, well behind Hank Greenberg who hit 44 in the AL. Kiner then led or tied for the major league lead in home runs each year for the end of the decade, twice reaching the 50-home run level. In 1947 Johnny Mize tied Kiner with 51 home runs, and then again tied Kiner for the major league lead with 40 in 1948. Two important milestones were reached during the decade. Jimmie Foxx finished the 1940 season with exactly 500 home runs, becoming the second player to achieve that total. He finished his career with 534. Mel Ott hit his 500th home run during the 1945 season. In total during the decade, 50 or more home runs were hit three times but 40 home runs were only reached six times.

The two best known hitters during the '40s were Williams and DiMaggio, despite both losing 3 years serving during the war. The players were almost traded for each other before the 1947 season. One could only imagine how the "short porch" in right field at Yankee Stadium would have helped Williams, while DiMaggio, whose power numbers were hurt by the vastness of left field in Yankee Stadium, may have thrived aiming for the Green Monster at Fenway Park. The third best hitter was Stan Musial. Converted from a pitcher to an outfielder, he was brought up the last two weeks of the season in 1941 and hit .426. He went on to lead all of baseball in hitting three times from 1943 to 1948.

Perhaps the most significant event of the 1940s occurred on April 15, 1947 when Jackie Robinson made his debut for the Brooklyn Dodgers, thus becoming the first African American player in the major leagues. Despite initial opposition from some of his teammates as well as opposing players, Robinson excelled, winning the first ever NL Rookie of the Year Award. He led the NL in batting with a .342 average in 1949. In July of 1947 Larry Doby joined the Cleveland Indians, breaking the color barrier in the AL.

The 1950s

The two leading home run hitters of the decade of the '50s were two teammates, Duke Snider and Gil Hodges. What is remarkable is that the 43 home runs that Snider hit in 1956 was the only year that one of them was the league leader in home runs. Snider had a remarkably consistent run of five seasons, 1953 to 1957, hitting at least 40 home runs each year, but a total of "only" 207. Gil Hodges was also consistent, reaching 40 home runs twice, 30 home runs four times, and 20 home runs the other 4 years. Eddie Mathews led the major leagues twice, with 47 in 1953 and 46 in 1959, and Ralph Kiner extended his major league leading home run streak to 6 years.

Ted Williams, though aging and often injured, was still a dominant player until his retirement in 1960. Though again missing most of two seasons due to military service during the Korean War, he hit 227 home runs during the '50s. He hit .366 for the decade, including hitting .388 in 1957, the year he turned 39.

Joe DiMaggio played his last games in 1951. He retired with 361 total home runs, which at the time was fifth highest. The Yankees had a dominant player for over 30 years: Ruth to Gehrig to DiMaggio. The next star, Mickey Mantle, made his debut in 1951. He started slowly, but improved every year, having his all-around best year in 1956, when he won the major league triple crown with 52 home runs, 130 RBIs, and a .353 batting average. He also led the major leagues in runs scored, with 132.

Stan Musial continued to be one of the best hitters in baseball, leading the NL in batting each of 1950, '51, and '52 and then again in 1957 when at the age of 36 he had a .355 average. When he retired in 1963, his 463 home runs were sixth highest of all time.

Two other all-time greats made their major league debuts in the '50s. Like Mantle, Willie Mays began his major league career in 1951 and started slowly. He missed the 1953 season because of military service, but when he returned in 1954 his career took off. He led the majors with a .345 batting average and hit 41 home runs. The next year he led the major leagues for the first time with 51 home runs. Hank Aaron joined the Milwaukee Braves in 1954. He led the NL in batting twice, .328 in 1956 and .355 in 1959, and led the major leagues in home runs for the first time in 1957 with 44.

During the '50s Mantle and Mays were the only two players to hit 50 home runs, but the 40-home run level was reached 32 times.

The most significant occurrence of the decade occurred in 1958, when the Dodgers and the Giants left New York for Los Angeles and San Francisco, respectively. Before this teams had been stable, with only a few relocations. In 1953 the Boston Braves moved to Milwaukee; in 1954 the St. Louis Browns moved to Baltimore and became the Orioles; in 1955 the Athletics moved west, but only from Philadelphia to Kansas City. Baseball was restricted to the northeast quarter of the country. Now MLB spanned the country. Train travel became more frequently replaced by air travel.

The 1960s

Ted Williams' last year in baseball was 1960, when at the age of 41 he hit 29 home runs, including one his last at-bat, and batted .316. He finished with 521 home runs, which moved him into third place at the time, behind Ruth and Foxx. This was the first time since 1937 that the top four list of home run hitters were not Ruth, Foxx, Ott, and Gehrig. Williams' batting average of .344 is sixth all-time.

The 1960s was really a decade of change for MLB. Since 1903 each league had 8 teams and played a 154-game schedule, with each team playing the other 7 teams in their league 22 times. The relocation of the Dodgers and the Giants to California sparked interest in forming a new major league, with teams in markets that were without teams. To counter this new league, the NL and then the AL proposed expansion. In 1961 the AL acted first, placing a new team in Los Angeles, the Angels, relocating the Washington Senators from D.C. to Minneapolis and renaming the club the Minnesota Twins, and adding an expansion team in Washington D.C., taking the name Senators. The AL now had 10 teams; the 154-game schedule no longer provided a balanced schedule, so the number of games was increased to 162, allowing each pair of teams to meet 18 times. The NL did not expand until the next year. To make up for the defections of the Dodgers and the Giants to the west coast, New York was awarded a new team, the Mets, who were to play in Flushing Meadows, Queens. The other new team was the Houston Colt .45s, later renamed the Astros, bringing baseball to the state of Texas. The NL, following the AL lead, also increased the number of games to 162.

The question for the 1961 season was, with the extra games could anyone step up and challenge Babe Ruth's record of 60 home runs? A pair of teammates, Mickey Mantle and Roger Maris, accepted the challenge. The two went back and forth with the home run lead. Mantle started quickly, and Maris was slow out of the gate, but quickly got hot. By the first All-Star break, July 10, Maris led Mantle 33 to 29. Shortly after, the Commissioner of Baseball, Ford Frick, who had been a friend of Babe Ruth, declared that Ruth's record of 60 home runs needed to be surpassed within 154 games to be the official record. If the 61st home run occurred during the last eight games of the season, then the total would have an asterisk indicating it was done using the extra games. A month later, on August 13 they were tied at 45 home runs. By September 3, both players had reached the 50-home run plateau, Maris with 53 and Mantle 50. After a series with the Cleveland Indians, Maris stood at 56, Mantle 53. Their total of 109 surpassed the total of 107 hit by Ruth and Gehrig in 1927, the record by teammates. Unfortunately for Mantle, his body began to break down; he only hit one more home run the rest of the season. Maris began to really feel the pressure, so much so his hair began to fall out. The press and the fans were against him, and Mantle—who had been maligned before 1961—became the fan favorite during the home run chase. Maris did not hit his 60th home run until game 158. While he took four more games than Ruth, he actually had fewer at-bats. He hit his 61st home run in the final game of the season. Neither Maris nor Mantle came close to their totals the rest of their careers. Mantle, who hung up his cleats in 1968, finished with 536 home runs, which was third highest at the time.

During the decade of the 1960s Harmon Killebrew hit 393 home runs, and led the AL in home runs five times, twice hitting 49, the other three 48, 45, and 44. Hank Aaron hit 40 home runs five times, winning the NL title three of those years. Aaron was a remarkable model of consistency. Over 23 seasons he hit 755 home runs, but the maximum any year was 47, and from 1957 to 1973 at least 30 home runs per year, except for 2 years. He hit his 500th home run during the 1968 season. Willie Mays also led the NL in home runs three times, hitting 52 in 1965, the year he passed the 500-home run mark. Mays continued to hit during the decade, finishing 1969 with exactly 600 home runs, obviously the second player to hit that many. Frank Robinson, who won the MVP with Cincinnati in 1961, was traded to the Baltimore Orioles in 1966, led the majors with 49 home runs, won the triple crown, and became the first player to win the MVP in both leagues. Eddie Mathews hit his 500th home run in 1967, finishing with 512.

In addition to Maris's 61, Mantle's 54, and Mays's 52, there were 31 other times that players hit at least 40 home runs. Five players reached 500 home runs for their career.

The New York Mets lost a record 120 games in their initial season and finished no higher than ninth place the next 6 years, but shocked the baseball world by beating the Baltimore Orioles 4 games to 1 in the 1969 World Series.

There were more relocations and expansion by the end of the decade. In 1966 the Braves moved from Milwaukee to Atlanta, the first team to play in the southeast. In 1968 the Athletics again relocated from Kansas City to Oakland. This move caused baseball to move up their expansion plans, adding two teams to each league: the Kansas City Royals and the Seattle Pilots in the AL, and the San Diego Padres and the Montreal Expos in the NL. There were now five teams in California and the first team outside of the United States. The expansion of teams did not increase the number of games played but resulted in each league being divided into two divisions and adding a new round of playoffs to determine the pennant winners.

The 1970s

The most significant event of the 1970s occurred in 1974. Hank Aaron, who reached 600 home runs in 1971, the third player to do so, ended the 1973 season with 713 home runs, one behind the most sacred record in baseball, the 714 hit by Babe Ruth. Aaron was under intense pressure in his pursuit of the record, even more than Maris in 1961, as he was bombarded with hate mail and received numerous death threats. The 6 months during the offseason had to seem like a lifetime. Aaron's team, the Atlanta Braves (they had moved from Milwaukee in 1966), wanted him to sit out the first series of the year in Cincinnati so that he could break the record at home, but

Commissioner Bowie Kuhn ruled he had to play two games in the opening series. It took him one swing to tie Ruth at 714, he sat out the second game, and he went hitless in the third. On April 8, before the largest crowd in Braves history, Aaron connected in his second at-bat against Al Downing and the record was his. Aaron would hit 40 more home runs before ending his career in 1976 with 755. He also holds the all-time record with 2297 RBIs, again surpassing Ruth who had finished with 2213.

In 1970 Johnny Bench led the majors in home runs with 45, one of six players to reach 40 that year. Over the rest of the decade 40 home runs were hit only 14 more times, the most by George Foster with 52 in 1977. There were 3 years, 1974 to 1976, when no player reached 40 and one additional year, 1972, when Bench hit exactly 40. Willie Stargell, who hit the most home runs during the decade, led the majors twice, in 1971 and 1972. Reggie Jackson, the second leading home run hitter of the decade with 292, led the AL in home runs twice, though he never reached the 40-home run level. Mike Schmidt led the majors each of 1974–1976, but never hit more than 38 home runs.

Four players reached the 500-home run list during the 1970s—Ernie Banks who wound up with 512, tying Eddie Mathews; Willie McCovey who hit 521, tying Ted Williams; Harmon Killebrew who hit 573, which at that point was fifth all-time; and Frank Robinson who hit 586, which was fourth. The top five home run hitters remained the same until the year 2000.

The AL expanded again in 1977, adding the Toronto Blue Jays (now both leagues had a team in Canada) and the Seattle Mariners.

The 1980s

Mike Schmidt won five NL home run titles during the 1980s, to go along with the three titles he had won in the previous decade. Dale Murphy won two titles, one of which he shared with Schmidt in 1984. Eddie Murray won the AL home run title in 1981 with a total of 22, the low number attributed to the strike-shortened season. In 1983 he hit 33, his career best total, but he still managed to hit over 500 home runs in his career. In 1987 Mark McGwire set a rookie record by hitting 49 home runs, winning the AL rookie of the year. In 1989 Kevin Mitchell led the majors with 47 home runs, which was 25 more than his previous high mark.

Overall power hitting was down in the major leagues. Only 13 times were 40 or more home runs hit during the 1970s, the lowest total number of any decade since the 1920s. There were several factors that contributed to this decline. Artificial turf began to be installed in ballparks. Baseballs bounced higher and traveled faster. Teams changed their lineups to take advantage of the new turf and focused on speed. Another factor was the rise of relief pitching, especially the role of the closer. Goose Gossage, Jesse Orosco, and Bruce Sutter were three of the best who shut down batters at the end of games. Now batters were facing fresh pitchers at the end of games.

There were two more additions to the 500-home run club. Reggie Jackson hit his 500th in 1984 and retired with 563; Mike Schmidt joined the club in 1987 and had 548 by the end of his career. When they retired, they stood in sixth and seventh place, respectively, on the all-time list, sandwiched between Killebrew and Mantle.

As a final and fitting end to the decade, Pete Rose, who is the all-time hit leader with 4256, was banned from baseball for allegedly betting on baseball games while he was a player and later a manager.

Conclusion

From 1960 through 1989 there were 64 times that a player hit at least 40 home runs in a season. Three of them were in the '50s and one in the '60s (Roger Maris, who hit 61). Contrast that to just the 1990s, when at least 40 home runs were hit over 70 times, including 8 times in the 50s, 3 times in the 60s, and once in the 70s. By the end of the 1989 season the top 11 home run hitters of all time were Aaron (755), Ruth (714), Mays (660), Robinson (586), Killebrew (573), Jackson (563), Schmidt (548), Mantle (536), Foxx (534), Williams (521), and McCovey (521). They remained that way through the end of the 1998 season. By 2003 only Aaron, Ruth, Mays, and Robinson remained on the list. These numbers are indicative of the problems that baseball purists have with the records produced during the steroid era.

References

www.mlb.com
en.wikipedia.org/wiki/List_of_Major_League_Baseball_progressive_career_
    home_runs_leaders
www.baseball-reference.com/leaders/HR_top_ten.shtml
thisgreatgame.com/baseball-lists-top-hitters
www.baseball-almanac.com

Bill James and the genesis of sabermetrics

4

Chapter outline

The epiphany occurred on an ordinary summer day. According to Alan Schwarz in his book *The Numbers Game*, Bill James was wlking to his job as a night watchman at the Stokely–Van Camp canning plant in Lawrence, Kansas. He chatted with a neighbor who was listening to a KC Royals game on the radio, and she would pause occasionally from gardening to write something down. When asked, she told him that she was keeping score of the game. In that moment, James realized that there were others fascinated as deeply as he about baseball numbers (Schwarz, 111).

He tried (and succeeded) in forging a career working with baseball statistics: and the process forever changed the way we view the game. He published 12 annual editions of *Baseball Abstract*, the last six of which were marketed and sold widely. In his 1984 *Baseball Abstract*, he wrote, "I am something of a baseball agnostic; I make it a point never to believe anything just because it is widely known to be so." That attitude was evident in one of the articles he wrote for the monthly "Baseball Digest" in those early years, which tackled the myth of fielding percentage.

Range factor

Before Bill James and baseball's statistical revolution, fielding prowess was quantified by using the fielding percentage, which has the formula

$$\text{FLD} = \frac{\text{PO} + \text{A}}{\text{TC}} = \frac{\text{PO} + \text{A}}{\text{PO} + \text{A} + \text{E}}$$

Sabermetrics. DOI: http://dx.doi.org/10.1016/B978-0-12-822345-1.00004-0

PO = putouts, A = assists, E = errors, TC = total chances.

It has become a fairly consensus position that former Atlanta Braves center fielder Andruw Jones is one of the top defensive center fielders in baseball history. In 2006, before the outfield Gold Glove Awards were split by an outfield position, two center fielders received the award, Jones of the Braves and Mets CF Carlos Beltrán. Their numbers that season were as follows (all data in this chapter come from www.baseballreference.com):

2006	Putouts	Assists	Errors
Carlos Beltrán	357	13	2
Andruw Jones	378	4	2

Using the above formula, Jones' fielding percentage was 0.9948, while Beltrán's was 0.9946, virtually identical. Is there any other way to measure how good a defensive season each player had?

Bill James observed that the traditional way of judging fielding prowess by the proportion of successful chances to unsuccessful chances was insufficient. He felt that there was more to the story. Suppose one player made a successful play on every ball hit to him, say, 100 total chances with no errors. Comparing him with another player who successfully handled 200 total chances but made errors on 10 others. Devoid of any situational context, it would seem that the second player is doing more to prevent the other team from winning than the first player. He introduced a statistic that he called range factor (RF) to give a better picture of a player's defensive contribution. As mentioned in Chapter 3, the statistic was first introduced by Henry Chadwick in 1872 and resurrected by James:

$$RF = \frac{PO + A}{G}.$$

For Beltran in 2006:

$$RF = \frac{PO + A}{G} = \frac{357 + 13}{136} = 2.74,$$

while for Jones, it's: $RF = \frac{PO + A}{G} = \frac{378 + 4}{153} = 2.51$

Thus, Beltran made more successful defensive plays per game than Jones in 2006.

In 1976, Garry Maddox patrolled center field (CF) CF for the Philadelphia Phillies, while the LA. Dodgers had Dusty Baker getting the most playing time at CF. Their respective fielding numbers at CF are:

1976	Putouts	Assists	Errors		g	fpct
Maddox	441	10	5		144	0.989
Baker	209	3	1		83	0.995

So, Baker had a 0.995 fielding percentage, making a single error, while Maddox's fielding percentage was at 0.989, charged with five errors. For Baker:

$$RF = \frac{PO + A}{G} = \frac{209 + 3}{83} = 2.57 \cdot$$

Maddox's was $RF = \dfrac{PO + A}{G} = \dfrac{441 + 10}{144} = 3.17.$

Now, one could argue that the RF for Maddox is a byproduct of playing more games. However, consider total chances (TC); Maddox had TC = 451, while Baker had TC = 212. So, we can give Baker 60 more games, and keep his RF the same; this would add $60 \times 2.57 = 154$ TC to his tally, giving him a theoretical TC = 366. He still would be 85 TC behind the Gold Glove winner, nicknamed "The Secretary of Defense"; whom it was said, "Two-thirds of the earth is covered by water, the rest is covered by Garry Maddox."

As with any mathematical model, the more data put into it, the better the measurement. For example, as more granular data becomes freely available, there has been a refinement to the RF statistic over the years, namely, the RF per nine innings (RF/9). This helps to give credit to the late-inning defensive replacement, not "charging" him for a whole game when he only played an inning or two: $RF/9 = 9\left(\dfrac{PO + A}{I}\right).$

If we put their RF on this nine-inning basis, we obtain the following:

1976	po	a	e	g	fpct	rf	CF inn	rf/9
Maddox	441	10	5	144	0.989	3.17	1240	3.27
Baker	209	3	1	83	0.995	2.57	729	2.62

Thus, on a nine-inning basis in CF, Maddox made 3.27 successful plays while Baker made 2.62.

Exercises

1. In 1978 the National League was blessed with four outstanding shortstops: Dave Concepción of the Cincinnati Reds, Larry Bowa of the Philadelphia Phillies, Garry Templeton of the St. Louis Cardinals, and rookie Ozzie Smith of the San Diego Padres. Given their putouts,

assists, errors, games, and inning totals (see chart), calculate their fielding percentages, RF, and RF per nine innings.

	PO	Assists	Errors	Games	Inning Totals
Concepcion	255	459	23	152	1307.33
Bowa	224	502	10	156	1350.66
Templeton	285	523	40	155	1353.66
Smith	264	548	25	159	1327.00

2. Larry Bowa was awarded the Gold Glove for NL shortstops in 1978. Based on your work in problem 1, do you feel that it was the right choice?

Hall of Fame, part 1

It is common folklore that the reason we study sabermetrics is the existence of the Baseball Hall of Fame. Over the years, James has produced various Hall of Fame Monitors that use statistics to try to predict whether a player would ultimately be inducted into the Hall. His success with the annual abstracts and two editions of the *Historical Baseball Abstract*, led to four other books: *The Politics of Glory* or *Whatever Happened to the Hall of Fame?*, *The Neyer/James Guide to Pitching* (co-authored with Rob Neyer), *The Bill James Guide to Baseball Managers from 1870 to Today* (with "today" being 1997), and *Win Shares*.

In his *1983 Baseball Abstract*, his second mass-produced abstract, James reprinted an article that he had written in his *1980 Baseball Abstract*, entitled "What Does It Take? Discerning the De Facto Standards of the Hall of Fame." He provides formulas to determine whether a pitcher or outfielder with certain career accomplishments is likely to make the Hall of Fame; however, he is very clear to write, "I am not in the least talking about what Hall of Fame standards *should be*. I am talking about what they *are*." In other words, this should not be interpreted as his opinion on who should be enshrined, but more of a system for determining who might be enshrined.

The premise of this particular study, which he called the Hall of Fame Prediction System (HOFPS), was designed in such a way that an outfielder who received 100 points or more would be in the Hall of Fame, while a score below that would mean that the player was not enshrined.

He had no pretense that the system made sense; the system awards eight points for each season that an outfielder batted above .300, while awarding three points for each 100 runs batted in (RBIs) season. A situation he termed "asinine," as a statement of the relative values of these events.

The system for pitchers at that time was as:

1. Count 3 points for each season of 15 to 19 wins.
2. Count 10 points for each season of 20 to 29 wins (although).
   a. 20–20 seasons should count as 15–win seasons.
   b. Seasons should not be counted in more than 1 category; for example, 20–win seasons do not receive both 3 points and 10 points.
3. Count 1 win for each win in a season of 30+.
4. Count 1 point for every 3 wins between 150 to 235, and then 1 point for each win above 235.
5. Count 2 points for each World Series start.
6. Count 3 points for each World Series win [World Series points not to exceed 30].
7. Count 5 points for each season leading the league in ERA.
8. Count 5 points for each season leading the league in strikeouts.
9. Count 1 point for each 0.01 the pitcher's career winning pct. is above 0.500.

Applying this system to Dodgers pitcher Don Drysdale whose career record was 209 to 166 with a 2.95 ERA, yields the following.

1. Count 3 points for each season of 15 to 19 wins; [15 points—1957, 1959, 1960, 1963, 1964].
2. Count 10 points for each season of 20 to 29: [20 points—1962, 1965].
3. Count 1 win for each win in a season of 30+: [no points].
4. Count 1 point for every 3 wins between 150 to 235, and then 1 point for each win above 235: [20 points—209 −150 = 59, 59/3 round up to 20].
5. Count 2 points for each World Series start: [12 points—6 World Series starts].
6. Count 3 points for each World Series win [World Series points not to exceed 30]: [9 points—3 World Series wins].
7. Count 5 points for each season leading the league in ERA: [no points].
8. Count 5 points for each season leading the league in strikeouts: [15 points—led in 1959, 1960 and 1962].
9. Count 1 point for each 0.01 the pitcher's career winning pct. is above 0.500: [49 points—0.549 career winning percentage].

His score is 114. Drysdale was elected to the Baseball Hall of Fame in 1984, a year after the Abstract published the system.

Now, we apply this system to Mets (among others) pitcher Jerry Koosman. His career record was 222 to 209 and his ERA was 3.36. Here is how Koosman scores:

1. Count 3 points for each season of 15 to 19 wins: [12 points—1968, 1969, 1974, 1980].
   a. Count 10 points for each season of 20 to 29 wins [20 points—1976, 1979].
2. Count 1 win for each win in a season of 30+: [no points].
3. Count 1 point for every 3 wins between 150 and 235, and then 1 point for each win above 235: [24 points—222 −150 = 72, 72/3 = 24].
4. Count 2 points for each World Series start: [8 points—4 World Series starts].
5. Count 3 points for each World Series win [World Series points not to exceed 30]; [9 points—3 World Series wins].
6. Count 5 points for each season leading the league in ERA; [no points].
7. Count 5 points for each season leading the league in strikeouts; [no points].
8. Count 1 point for each 0.01 the pitcher's career winning pct. is above 0.500: [15 points—0.515 career winning percentage].

His score is 88. Koosman was dropped off the writers' ballot after he received only 4 votes in his first year of eligibility.

Exercises

1. Apply this system to the career statistics of pitcher Luis Tiant. Is he enshrined in the Hall of Fame? Comment on whether or not you agree with this state of affairs.
2. Apply this system to the career statistics of pitcher Jim Bunning. Is he enshrined in the Hall of Fame? Comment on whether or not you agree with this state of affairs.

Hall of Fame, part 2

Also in the 1983 Abstract, James gave a HOFPS for outfielders. He hypothesized that the system would work okay for first basemen and possibly third basemen. The system is:

1. Count 8 points for each season of batting .300 in 100 or more games (maximum of 60 points).
2. Count 15 points for a lifetime batting average of .315 or better in 1000 games;
3. Count 3 points for each 100+ RBIs season.
4. Count 8 points for each season of 200+ hits.
   a. Count 4 points for every season leading the league in stolen bases.

   b. Count 5 points for every season leading the league in RBIs.
   c. Count 8 points for every season leading the league in HR.
   d. Count 12 points for every season leading the league in batting average.
5. Count 1 point for each World Series game played, at a maximum of 18.
6. Count 10 points if the player had 3000+ career hits.
7. Count 10 points if the player had 400+ career HR.

We apply this system to 1940s and 1950s slugging outfielder Ralph Kiner, who played for the Pittsburgh Pirates (among others). His career numbers were 369 HR/1015 RBI/0.279 batting average. The system scores him as:

1. Count 8 points for each season of batting .300 in 100 or more games (maximum of 60 points): [24; 1947, 1949 and 1951].
2. Count 15 points for a lifetime batting average of .315 or better in 1000 games: [no points].
3. Count 3 points for each 100+ RBIs season: [18; 1947, 1948, 1949, 1950, 1951, 1953].
4. Count 8 points for each season of 200+ hits; [no points].
   a. Count 4 points for every season leading the league in stolen bases; [no points].
   b. Count 5 points for every season leading the league in RBIs; [5; 1949].
   c. Count 8 points for every season leading the league in HR; [56; 1946 to 1952, inclusive].
   d. Count 12 points for every season leading the league in Batting Average; [no points].
5. Count 1 point for each World Series game played, at a maximum of 18; [no points].
6. Count 10 points if the player had 3000+ career hits; [no points].
7. Count 10 points if the player had 400+ career HR; [no points].

This gives Kiner 103 points. He was inducted into the Hall of Fame in 1975.
Irish Meusel was an outfielder for the New York Giants in the 1910s and 1920s. His career numbers were 106/819/.310. The system scores him as:

1. Count 8 points for each season of batting .300 in 100 or more games (maximum of 60 points); [48; 1919, 1920, 1921, 1922, 1924 and 1925].
2. Count 15 points for a lifetime batting average of .315 or better in 1000 games; [no points].

3. Count 3 points for each 100+ RBIs season: [12; 1922 to 25].
4. Count 8 points for each season of 200+ hits: [16; 1921, 1922].
   a. Count 4 points for every season leading the league in stolen bases; [no points].
   b. Count 5 points for every season leading the league in RBIs: [5; 1923].
   c. Count 8 points for every season leading the league in HR: [no points].
   d. Count 12 points for every season leading the league in batting average [no points].
5. Count 1 point for each World Series game played, at a maximum of 18 [18; 23 games played from 1921 to 1924].
6. Count 10 points if the player had 3000+ career hits [no points].
7. Count 10 points if the player had 400+ career HR [no points].

Irish Meusel scores 99 points and has yet to be inducted into the Hall of Fame.

Exercises

1. Apply this system to the career statistics of outfielder Dixie Walker. Is he enshrined in the Hall of Fame? Comment on whether or not you agree with this state of affairs.
2. Apply this system to the career statistics of outfielder Al Kaline. Is he enshrined in the Hall of Fame? Comment on whether or not you agree with this state of affairs.

Hall of Fame, part 3

In his 1995 book *The Politics of Glory*, James presents Hall of Fame standards phrased as: What level of statistical performance truly constitutes a Hall of Fame standard? (Whatever Happened to the Hall of Fame, 174–176). Unlike the HOFPS, which relied on some seasonal data, the standards are based entirely on career total, and, in some cases, proportional data. It is simply a measure of where the players compare to each other and, perhaps, how players who have not been inducted compare to those who have been. In the standards, the average Hall of Famer will score a 50 in this system; therefore, while there is no number for which we could say that a player should definitely be inducted or not, James claims that a candidate who meets at least 35% of the standards is a viable Hall of Fame candidate. The standards for everyday players has 11 separate components and the total score is the percentage of standards that he meets. For everyday players, it is:

1. For career hits, count 1 point for each 150 hits above 1500 (maximum of 10 points).

2. For batting average, count 1 point for each 0.005 above 0.275 (maximum of 9 points), with an extra point if the career average is 0.300 or above.
3. For runs scored, count 1 point for each 100 runs above 900 (maximum of 8 points), an additional point if the player scored at least 1 run for every 2 games played, and another point if the player scored at least 1 run for every 1.55 games played.
4. For RBI, count 1 point for each 100 RBIs above 800 (maximum of 8 points), an additional point if the player drove in at least 1 run for every 2 games played, and another point if the player drove in at least 1 run for every 1.67 games played.
5. For slugging percentage, count 1 point for each 0.025 above 0.300 (maximum of 10 points).
6. For on base average, count 1 point for each 0.010 above 0.275 (maximum of 10 points).
7. For HR, count 1 point for every 200 HR (to a maximum of 3 points), 1 point if 10% to 19% of career hits are HR, and 2 points if at least 20% of career hits are HR.
8. For extra base hits, count 1 point if at least 500, and an additional point for every 200 above that (maximum of 5 points).
9. For SB, 1 point for every 100 SB (maximum of 5 points).
10. For walks, count 1 point if at least 500, and an additional point for every 200 above that (maximum of 5 points).
11. Defensive position for 1000 games played at each.
    a. 20 points for catcher
    b. 16 points for SS
    c. 14 points for 2B
    d. 13 points for 3B
    e. 12 points for CF
    f. 6 points for RF
    g. 3 points for LF
    h. 1 point for 1B

Note that the positional difficulty scores are another Bill James wrinkle; he often referred to a defensive spectrum, in which positional difficulty from a defensive standpoint is considered (1982 Abstract). Leftmost on this spectrum is first base, followed by, from left to right, LF, RF, 3B, CF, 2B, and SS. Catchers, because of the extreme physical demands of the position, have their own considerations. As one moves from left to right, defensive responsibility increases and, conversely, offensive expectation decreases. Also, James asserts that as players age, they tend to move from right to left on the spectrum (1982 Abstract). In this system, there are more points awarded for players whose primary position is considered more challenging defensively. This innovation also is a component in calculation of the more modern Wins Above Replacement (WAR) statistic. There are no points for players whose primary position was DH.

Now, we apply these standards to Ralph Kiner's career:

1. For career hits, count 1 point for each 150 hits above 1500 (maximum of 10 points) [no points].
2. For batting average, count 1 point for each 0.005 above 0.275 (maximum of 9 points), with an extra point if the career average is 0.300 or above; [no points].
3. For runs scored, count 1 point for each 100 runs above 900 (maximum of 8 points), an additional point if the player scored at least 1 run for every 2 games played, and another point if the player scored at least 1 run for every 1.55 games played: [971 runs scored, 1472 games played, so 2 points].
4. For RBI, count 1 point for each 100 RBIs above 800 (maximum of 8 points), an additional point if the player drove in at least 1 run for every 2 games played, and another point if the player drove in at least 1 run for every 1.67 games played: [1015 RBIs: 1472 games played, 2 points].
5. For slugging percentage, count 1 point for each 0.025 above 0.300 (maximum of 10 points): [0.548 SLG: 9 points].
6. For on base average, count 1 point for each 0.010 above 0.275 (maximum of 10 points) [0.398 OB: 9 points].
7. For HR, count 1 point for every 200 HR (to a maximum of 3 points), 1 point if 10% to 19% of career hits are HR, and 2 points if at least 20% of career hits are HR [369 HR, 1472 games played, 3 points].
8. For extra base hits, count 1 point if at least 500, and an additional point for every 200 above that (maximum of 5 points) [624 XBH: 1 point].
9. For SB, 1 point for every 100 SB (maximum of 5 points): [no points].
10. For walks, count 1 point if at least 500, and an additional point for every 200 above that (maximum of 5 points): [1011 BB: 3 points].
11. Defensive position: [3 points for LF].

So Kiner scores 32 points by these standards; according to James, it means that he has met 32% of the Hall of Fame Standards.

Irish Meusel is not as close to being a Hall of Famer in this system as he is in the HOFPS. His career scores in this way:

1. For career hits, count 1 point for each 150 hits above 1500 (maximum of 10 points) [no points].
2. For batting average, count 1 point for each 0.005 above 0.275 (maximum of 9 points), with an extra point if the career average is 0.300 or above [0.310 average, 8 points].

3. For runs scored, count 1 point for each 100 runs above 900 (maximum of 8 points), an additional point if the player scored at least 1 run for every 2 games played, and another point if the player scored at least 1 run for every 1.55 games played: [701 runs scored; 1289 games played, so 1 point].

4. For RBI, count 1 point for each 100 RBIs above 800 (maximum of 8 points), an additional point if the player drove in at least 1 run for every 2 games played, and another point if the player drove in at least 1 run for every 1.67 games played: [819 RBIs; 1289 games played, so 2 points].

5. For slugging percentage, count 1 point for each 0.025 above 0.300 (maximum of 10 points): [0.464 SLG: 6 points].

6. For on base average, count 1 point for each 0.010 above 0.275 (maximum of 10 points): [0.348 OB: 4 points].

7. For HR, count 1 point for every 200 HR (to a maximum of 3 points), 1 point if 10% to 19% of career hits are HR, and 2 points if at least 20% of career hits are HR: [113 HR: 0 points].

8. For extra base hits, count 1 point if at least 500, and an additional point for every 200 above that (maximum of 5 points): [449 XBH: 0 points].

9. For SB, 1 point for every 100 SB (maximum of 5 points): [113 SB; 1 point].

10. For walks, count 1 point if at least 500, and an additional point for every 200 above that (maximum of 5 points): [269 BB; 0 points].

    a. Defensive position: [3 points for LF]

By this method, Irish Meusel scores 25 points and thus meets 25% of the Hall of Fame Standards.

Exercises

1. Apply the Standards to the career statistics of outfielder Dixie Walker. Comment on whether or not you agree with his Hall of Fame status.

2. Apply the Standards to the career statistics of outfielder Al Kaline. Is he enshrined in the Hall of Fame? Comment on whether or not you agree with his Hall of Fame status.

Hall of Fame, part 4

James provides 10 separate Standards for pitchers. Note that there are no Standards yet for relief pitchers, so these only should be applied to starting pitchers:

1. For career wins, count 1 point for each 10 wins above 100 (maximum of 25 points).

2. For winning percentage, count 1 point for each 0.013 above 0.500 (maximum of 15 points).
3. For games over 0.500 (wins minus losses), count 1 point for each 20 (maximum of 10 points).
4. For career ERA, count 1 point for each 0.2 below 4.00 (maximum of 10 points).
5. For strikeouts, count 1 point for each 200 above 1000 (maximum of 10 points).
6. For walks per 9 innings, count 1 point for each 0.3 below 4 walks per game (maximum of 10 points).
7. For hits per 9 innings, count 1 point for each 0.3 below 10 hits per game (maximum of 10 points).
8. For IP, count 1 point for each 1000 IP above 1000 (maximum of 5 points).
9. For CG, 1 point if 200 CG, 2 points if 305 CG, and 3 points if 500 CG.
10. For shutouts (SHO), count 1 point if 30 SHO, 2 points is 60 SHO.

For Drysdale, the standards are:

1. For career wins, count 1 point for each 10 wins above 100 (maximum of 25 points): [209 wins; 10 points].
2. For winning percentage, count 1 point for each 0.013 above 0.500 (maximum of 15 points): [0.557 winning pct.; 4 points].
3. For games over 0.500 (wins minus losses), count 1 point for each 20 (maximum of 10 points): [209 to 166 = 43; 2 points].
4. For career ERA, count 1 point for each 0.2 below 4.00 (maximum of 10 points): [2.95 ERA; 5 points].
5. For strikeouts, count 1 point for each 200 above 1000 (maximum of 10 points): [2486 K; 7 points].
6. For walks per 9 innings, count 1 point for each 0.3 below 4 walks per game (maximum of 10 points): [2.2 BB/9IP; 6 points].
7. For hits per 9 innings, count 1 point for each 0.3 below 10 hits per game (maximum of 10 points): [8.1 H/9IP; 6 points].
8. For IP, count 1 point for each 1000 IP above 1000 (maximum of 5 points): [3432 IP; 2 points].
9. For CG, 1 point if 200 CG, two points if 305 CG, and 3 points if 500 CG: [167 CG; no points].
10. For shutouts (SHO), count 1 point if 30 SHO, 2 points is 60 SHO: [49 SHO: 1 point].

Hence, Drysdale satisfies 43% of the Hall of Fame standards.
Jerry Koosman's score is:

1. For career wins, count 1 point for each 10 wins above 100 (maximum of 25 points): [222 wins; 12 points].
2. For winning percentage, count 1 point for each 0.013 above 0.500 (maximum of 15 points): [0.515 winning pct.; 1 point].
3. For games over 0.500 (wins minus losses), count 1 point for each 20 (maximum of 10 points: [222 to 209 = 13; 0 points].
4. For career ERA, count 1 point for each 0.2 below 4.00 (maximum of 10 points): [3.36 ERA; 3 points].
5. For strikeouts, count 1 point for each 200 above 1000 (maximum of 10 points): [2556 K; 7 points].
6. For walks per nine innings, count 1 point for each 0.3 below 4 walks per game (maximum of 10 points): [2.8 BB/9 IP; 4 points].
7. For hits per nine innings, count 1 point for each 0.3 below 10 hits per game (maximum of 10 points): [8.5 H/9IP; 5 points].
8. For IP, count 1 point for each 1000 IP above 1000 (maximum of 5 points): [3839 2/3 IP; 2 points].
9. For CG, 1 point if 200 CG, 2 points if 305 CG, and 3 points if 500 CG: [140 CG; no points].
10. For shutouts (SHO), count 1 point if 30 SHO, 2 points if 60 SHO: [33 SHO; 1 point].

Jerry Koosman meets 35% of the Standards, right at the point where Hall of Fame consideration becomes serious.

Exercises

1. Apply this system to the career statistics of pitcher Luis Tiant.
   Comment on whether or not you agree with his Hall of Fame status.
2. Apply this system to the career statistics of pitcher Jim Bunning.
   Comment on whether or not you agree with his Hall of Fame status.

Hall of Fame, part 5

Over the years, Bill James presented a Hall of Fame Monitor, which examined the progress to Hall of Fame candidacy of active players. In a combination of other methods presented previously, it awards points for some career statistics and some seasonal accomplishments. Once again, there are separate considerations for pitchers and hitters; one added feature is further distinction for players who play more difficult defensive positions. In addition, the pitching standards also can be applied to relievers. These were presented in *Whatever Happened to the Hall of Fame?* The system for everyday players is:

1. Batting average for seasons of 100 or more games:
   a. Fifteen points for each season above 0.399
   b. Five points for each season between 0.350 and 0.399 (inclusive)
   c. 2.5 points for each season between 0.300 and 0.349 (inclusive)
2. Seasonal totals:
   a. Five points for each season of 200 or more hits
   b. Three points for each season of 100 R or RBIs
3. Seasonal HR:
   a. Ten points for each season of 50 or more HR
   b. Four points for 40 to 49 HR
   c. Two points for each 30–39 HR season
4. Seasonal Doubles:
   a. Two points for each season of 45 or more doubles
   b. One point for each season of 35 to 44 doubles
5. Award Points:
   a. Eight points for an MVP season
   b. Three points for each All-Star game
   c. Two points for each season of Gold Glove at C, SS, or 2B
   d. One point for Rookie of the Year
6. World Series:
   a. Six points if a regular SS or C on a World Series champion
   b. Five points if regular CF or 2B
   c. Three points for 3B
   d. Two points for LF or RF
   e. One point for 1B
7. League Champion:
   a. Five points if a regular SS or C on a pennant winner
   b. Three points if regular CF or 2B
   c. One point for 3B
8. Division winner:
   a. Two points if a regular SS or C on a division winner
   b. One point if regular CF, 2B, or 3B
9. For every season leading the league in:
   a. Batting average (6 points)
   b. HR or RBI (4 points each)
   c. Runs (3 points)
   d. Hits or SB (2 points each)
   e. Doubles or Triples (1 point each)
10. Career base hits:
    a. Fifty points if 3500 or more career base hits
    b. Forty points for 3000–3499 H
    c. Fifteen points for 2500–2999 H
    d. Four points for 2000–2499 career H

11. Career HR:
    a. Thirty points if 600 or more career HR
    b. Twenty points for 500 to 599 HR
    c. Ten points for 400 to 499 HR
    d. Three points for 300 to 399 career HR
12. For career batting average:
    a. Twenty-four points if above 0.330
    b. Sixteen points for a career average from 0.315 to 0.329
    c. Eight points for 0.300 to 0.314
13. Career games played by position
    a. For catcher:
       i. Sixty points for 1800 or more games
       ii. Forty-five points for 1600 to 1799
       iii. Thirty points for 1400 to 1599
       iv. Fifteen points for 1200 to 1399
    b. For SS or 2B:
       i. Thirty points for 2100 or more games
       ii. Fifteen points for 1800 to 2099 games
    c. For 3B, 15 points for 2000 or more games
    d. An additional 15 points if total number of games at 2B, 3B, or SS is 2500 or greater
14. Fifteen points for 1500 games at 2B/SS/C and a career batting average of 0.275 or above.

If a player accumulates 130 or more points in this system, he is almost a certain Hall of Famer while 100 or more points indicates a likely Hall of Famer. Note that point number 8 renders the system somewhat unusable for pre-1969 baseball, while the All Star and Rookie of the Year considerations indicate it can only be used effectively after those awards were established.

Los Angeles Angels outfielder Mike Trout has put up career statistics that already have drawn comparison to some of the best players of all time, such as Frank Robinson and Mickey Mantle. Applying the Hall of Fame Monitor to his statistics through the 2020 season yields:

1. Batting average for seasons of 100 or more games:
   a. Fifteen points for each season above 0.399
   b. Five points for each season between 0.350 and 0.399 (inclusive)
   c. 2.5 points for each season between 0.300 and 0.349 (inclusive); ['12, '13, '14, '16, '17, '18: 12.5 points]
2. Seasonal totals:
   a. Five points for each season of 200 or more hits, 3 points for each season of 100 R or RBIs; [100 RBIs: '14, '16, '19 = 9 points; 100 R: '12-'19: 21 points].

3. Seasonal HR:
   a. Ten points for each season of 50 or more HR
   b. Four points for 40 to 49 HR, [1915, 1919 = 8 points]
   c. Two points for each 30 to 39 HR season; [1912, 1914, 1917, 1918 = 8 points]
4. Seasonal Doubles:
   a. Two points for each season of 45 or more doubles,
   b. One point for each season of 35 to 44 doubles; [1913–1916 = 4 points]
5. Award Points:
   a. |Eight points for an MVP season, [1914, 1916, 1919 = 24 points]
   b. Three points for each All-Star game, [1912–1919 = 24 points]
   c. Two points for each season of Gold Glove at C, SS, or 2B, [no points]
   d. One point for Rookie of the Year; [1912 = 1 point]
6. World Series [no points].
7. League Champion [no points].
8. Division Winner:
   a. Two points if a regular SS or C on a division winner
   b. One point if regular CF, 2B, or 3B; [1914 = 1 point]
9. For every season leading the league in:
   a. Batting average (6 points); [no points]
   b. HR or RBI (4 points each); [RBI 1914: 4 points]
   c. Runs (3 points); [1912, 1913, 1914, 1916 = 12 points]
   d. Hits or SB (2 points each); [SB 1912: 2 points]
   e. Doubles or triples (one point each); [no points]
10. Career Base Hits [no points]
11. Career HR:
    a. Thirty points if 600 or more career HR
    b. Twenty points for 500 to 599 HR
    c. Ten points for 400 to 499 HR
    d. Three points for 300 to 399 career HR; [302 HR = 3 points]
12. For career Batting Average:
    a. Twenty-four points if above 0.330
    b. Sixteen points for a career average from 0.315 to 0.329
    c. Eight points for 0.300 to 0.314; [0.304 = 8 points]
13. Career games played by position [no points].
14. Fifteen points for 1500 games at 2B/SS/C and a career batting average of 0.275 or above [no points].

Trout scores 141.5, well above James' stated level of Hall of Fame certainty. Now it remains to see where in the pantheon he will wind up as his career totals increase, and perhaps, further postseason involvement.

Exercises

1. Apply the Hall of Fame Monitor to the statistics of former Trout teammate Albert Pujols. Does his future enshrinement seem likely?
2. Apply the Hall of Fame Monitor to the statistics of veteran slugger Ryan Braun. Does his future enshrinement seem likely?

Hall of Fame, part 6

The Hall of Fame Monitor for pitchers is:

1. For wins in a season:
   a. Fifteen points for each season of 30 or more wins
   b. Ten points for 25 to 29
   c. Eight points for 23 to 24
   d. Six points for 20 to 22
   e. Four points for 18 to 19
   f. Two points for each season of 15 to 17 wins
2. For strikeouts in a season:
   a. Six points for seasons of 300 or more K
   b. Three points for each season between 250 and 299
   c. Two points for 200 to 249
3. For winning percentage in a season, 2 points for each season of 0.700 or better with 14 or more wins.
4. For ERA and seasons of either 50 or more games or 150 or more IP
   a. Four points for each season under 2
   b. One point for each season of an ERA between 2 and 2.99
5. For saves, in a season:
   a. Seven points for each season of 40 or more saves,
   b. Four points for 30 to 39
   c. One point for 20 to 29
6. One point for each no-hitter thrown.
7. For each season leading the league in
   a. ERA (2 points each)
   b. Games, wins, IP, winning pct., K, saves, and shutouts (1 point each)
   c. Complete games (0.5 point)
8. For career wins:
   a. 300 or more (35 points)
   b. 275 to 299 (25 points)
   c. 250 to 274 (20 points)
   d. 225 to 249 (15 points)
   e. 200 to 224 (10 points)
   f. 175 to 199 (8 points)
   g. 150 to 174 (5 points)

9. For career winning percentage, and a minimum of 200 decisions:
   a. 0.625 and above (8 points)
   b. 0.600 to 0.624 (5 points)
   c. 0.575 to 0.599 (3 points)
   d. 0.550 to 0.575 (1 point)
10. Award 10 points for a career ERA below 3.
11. For career saves:
    a. 300 or more saves (20 points)
    b. 200 to 299 saves (10 points)
12. For career games pitched:
    a. 1000 or more games (30 points)
    b. 850 to 999 games (20 points)
    c. 700 to 849 games, (10 points)
13. For career strikeouts:
    a. 4000 or more K (20 points)
    b. 3000 to 3999 K (10 points)
14. For World Series performance:
    a. Two points for each start
    b. One point for each relief appearance
    c. Two points for each win
15. Award 1 point for each playoff win
16. Season awards:
    a. Five points for the Cy Young Award
    b. Three points for each All Star game
    c. Two points for each Gold Glove
    d. One point for Rookie of the Year

We note that James did not make allowance for a pitcher winning an MVP awards. We would likely score them the same as for an everyday player, that is, 8 points. Since, up until now, we have not been able to apply the system to relievers, we apply the monitor to a current reliever, Aroldis Chapman. His numbers through 2020 yield the following results:

1. For wins in a season: [no points].
2. For strikeouts in a season: [no points].
3. For winning percentage in a season, 2 points for each season of 0.700 or better with 14 or more wins: [no points].
4. For ERA: and seasons of either 50 or more games or 150 or more IP.
   a. Four points for each season under 2, [1912, 1915, 1916—12 points]
   b. One point for each season of an ERA between 2 and 2.99 [1913, 1914, 1918, 1919—4 points]
5. For saves, in a season:
   a. Seven points for each season of 40 or more saves, [no points]

    b. Four points for 30–39 [1912, 1913, 1914, 1915, 1916, 1918, 1919—
    28 points]

    c. One point for 20–29 ['17—1 point]

6. One point for each no-hitter thrown: [no points].

7. For each season leading the league: [no points].

8. For career wins: [no points].

9. For career winning percentage and a minimum of 200 decisions: [no points].

10. Award 10 points for a career ERA below 3: [10 points].

11. For career saves:
    a. 300 or more saves (20 points)
    b. 200 to 299 saves (10 points) [10 points]

12. For career games pitched: [no points].

13. For career strikeouts: [no points].

14. For World Series performance:
    a. Two points for each start: [no points]
    b. One point for each relief appearance: [1916—5 G—5 points]
    c. Two points for each win: [1916—1 W—2 points]

15. Award 1 point for each playoff win: [1916—1 W, 1920—1 W—2 points].

16. Season awards: [as: 1912, 1913, 1914, 1915, 1918, 1919—
    18 points].

    Chapman totals 92 points in this system, and as he accumulates more saves and appearances, this total should increase to above the 100–point Hall of Fame viability threshold.

    Roy Halladay was one of the greatest starting pitchers for the first part of the 21st century, inducted posthumously into the Hall of Fame in 2019. The monitor scores him as:

1. For wins in a season:
   a. Fifteen points for each season of 30 or more wins: [no points]
   b. Ten points for 25 to 29: [no points]
   c. Eight points for 23 to 24: [no points]
   d. Six points for 20 to 22: ['03, '08, '10—18 points]
   e. Four points for 18 to 19: ['02, '11—8 points]
   f. Two points for each season of 15 to 17 wins: [1906, 1907, 1909—6 points]

2. For strikeouts in a season:
   a. Six points for seasons of 300 or more K: [no points]
   b. Three points for each season between 250 and 299: [no points]
   c. Two points for 200 to 249: [1903, 1908, 1909, 1910, 1911 to 10 points]

3. For winning percentage in a season, 2 points for each season of 0.700 or better with 14 or more wins: [1902, 1903, 1906, 1911 to 8 points].
4. For ERA, and seasons of either 50 or more games or 150 or more IP,
   a. Four points for each season under 2.
   b. One point for each season of an ERA between 2 and 2.99 [1902, 1905, 1908, 1909, 1910, 1911 to 6 points]
5. For saves in a season: [no points].
6. One point for each no-hitter thrown: [2 points—counting postseason]
7. For each season leading the league in: [no points].
8. For career wins:
   200 to 224 (10 points): [203 Wins—10 points].
9. For career winning percentage and a minimum of 200 decisions:
   0.625 and above (8 points): [0.659—8 points].
10. Award 10 points for a career ERA below 3: [no points].
11. For career saves: [no points].
12. For career games pitched: [no points].
13. For career strikeouts: [no points].
14. For World Series performance: [no points].
15. Award one point for each playoff win: [3 points].
16. Season awards:
    a. Five points for the Cy Young Award: [1903, 1910—10 points]
    b. Three points for each All Star game: [1902, 1903, 1905, 1906, 1908, 1909, 1910, 1911—24 points]
    c. Two points for each Gold Glove: [no points]
    d. One point for Rookie of the Year: [no points]

Exercises

1. Apply the Hall of Fame Monitor to the statistics of retired reliever Billy Wagner. Does his future enshrinement seem likely?
2. Apply the Hall of Fame Monitor to the statistics of current reliever Kenley Jansen. Does his future enshrinement seem likely?

Hall of Fame, part 7

The preceding sections on the Hall of Fame present models that grow increasingly more complex and increasingly more inclusive as they are refined. However, other than some rate statistics, the measures are somewhat traditional in their nature. Bill James' most recent foray into Hall of Fame consideration is what he calls the Hall of Fame Value Standard (https://sportsinfosolutionsblog.com/2018/12/10/the-hall-of-fame-value-standard/).

In Chapter 3, Eugene Reynolds outlines the complex computations that are used in calculating WAR and Win Shares (WS). WAR is a statistic that attempts to measure the wins for which a player can be credited over and above the number of wins with which a replacement level would be credited. The statistic puts pitchers and non-pitchers on the same scale and adjusts for era in a relative sense, so players can be compared across the entire spectrum. While there are several versions of the formula, we will use the one found in Baseball Reference. If we were to sum up the WAR totals for all players on a team, we ostensibly should obtain a number very close to the team's number of wins. For context, an average player may have a WAR of 2, an All Star would be at 5 or more, and an MVP candidate in excess of 8 (Babe Ruth in 1923 had a WAR of 14.1, the highest single-season total of all time for an everyday player) and the 100th highest season total for an everyday player. For lifetime numbers, Babe Ruth has the highest total at 182.5 (including his pitching career as well as his everyday career), while Carlos Beltran at 70.1 is in 100th place.

Meanwhile, Bill James developed WS as his own universal statistic. The number of available WS for a team in any season is three times its win total, and players are awarded shares of those wins based on virtually every aspect of play. For context, it is acknowledged that 20 WS in a season indicates All Star caliber play, 30 WS MVP caliber, and 40 WS is an all-time season. Babe Ruth's 1923 season is his career highest at 55. His career number is 756.

Because of the incredibly high number of innings pitched (and in most cases, throwing underhand from 45 feet), 19th-century pitchers amassed the highest seasonal totals for both Wins Above Replacement (WAR) and WS. The all-time highest single-season WAR total belongs to New York Metropolitan pitcher Tim Keefe, 20.2 in 1883. The fourth highest WAR was 19.4 by Providence Grays pitcher Old Hoss Radbourn with 19.4 in 1884. Keefe's WS for 1883 is 70, while Radbourn's 1884 WS tally is 89.

James' most recent evaluation of Hall of Fame viability comes in the form of the Hall of Fame Value Standard (HOFVS). He combines WAR and WS into a single statistic; according to James WS attempts to answer the question, "How many games did this player win for his team? WAR attempts to answer, "How many games did this player win for his team, above the number of wins that a scrub player out of the minors would have been able to win?" (Hall of Fame Value Standard).

James looked at a group of pre-2009 players. In his analysis, the standard deviation (in mathematics, a measure of the data's spread) for their career WS was 88.5, while the standard deviation for their career WAR is 22.1. Thus, he uses the following formula:

$$HOFVS = WS + 4 \times WAR.$$

According to his research (Hall of Fame Value Standard), an HOFVS of 560 or greater indicates a Hall of Famer, while very few players who score below 460 are enshrined.

Thus, for Babe Ruth, HOFVS = WS + 4 × WAR = 756 + (4 × 182.5) = 1486, the highest of any player in history.

Applying this metric to some of the players discussed in this chapter, we obtain:

	Win Shares	Wins Above Replacement	Hall of Fame Value Standard
Koosman	163	57	391
Drysdale	258	67	526
I Meusel	161	21.6	247.4
Kiner	242	47.9	433.6

Perennial Hall of Fame candidate Gil Hodges has HOFVS = 263 + (4 × × 43.9) = 438.6, which is higher than some Hall of Famers (including Kiner). James also discusses the candidacy of 19th-century shortstop Bill Dahlen, whose HOFVS = 394 + (4 × 75.4) = 695.6 renders him the best player not enshrined.

Finally, James makes an adjustment for catchers, who he feels get short-changed by both WAR and WS in terms of their defensive contributions. To rectify this, he adds 20% to their HOFVS score, which we will call HOFVS-C = 1.2 × HOFVS. For Johnny Bench, HOFVS = 356 + (4 × 75.2) = 686.6, far surpassing the baseline for enshrinement. With the adjustment, we have HOFVS-C = 1.2 × 686.6 = 788.16.

Exercise

Given the raw numbers for the catchers, compute HOFVS and HOFVS-C, and comment on their enshrinement or potential enshrinement.

	Win Shares	Wins Above Replacement	Hall of Fame Value Standard	hofvs-c
Thurman Munson	206	46		
Gary Carter	337	70.1		
Ted Simmons	315	50.3		

Works cited

Costa, Fr. Gabriel, Michael Huber, and John T. Saccoman. "Cumulative Home Run Frequency and the Recent Home Run Explosion." Baseball Research Journal, 2005, pp. 37–41.

James, Bill. The Bill James Baseball Abstract 1983. Ballentine Books, 1983.

James, Bill. The Bill James Baseball Abstract 1984. Ballentine Books, 1985.

James, Bill. Whatever Happened to the Hall of Fame: Baseball, Cooperstown, and the Politics of Glory. Simon and Schuster, 1995.

https://sportsinfosolutionsblog.com/2018/12/10/the-hall-of-fame-value-standard/

Love, Connor. "Cumulative Home Run Ratio and Today's Home-run Hitters." Society for American Baseball Research (SABR), sabr.org/latest/cumulative-home-run-ratio-and-todays-home-run-hitters/. Accessed on November 22, 2020.

https://en.wikipedia.org/wiki/Win_Shares

Schwarz, Alan. The Numbers Game: Baseball's Lifelong Fascination with Statistics. St. Martin's Press, 2004.

Rattling the sabermetrics 5

> *"You know what the difference between hittin' .250 and .300 is?*
> *It's 25 hits. Twenty-five hits in 500 at-bats is 50 points. OK? There's*
> *six months in a season. That's about 25 weeks. That means if you*
> *get just one extra flare a week. Just one. A gork. You get a ground*
> *ball, you get a ground ball with eyes. You get a dying quail. Just*
> *one more dying quail a week and you're in Yankee Stadium.*
> *You still don't know what I'm talkin' about, do ya? Get the hell outta here!"*
> —Crash Davis, catcher and minor league guru, "Bull Durham" (1988)

Chapter outline

Batter up!

The best way to enjoy a game of baseball is to watch it. The second best way is to listen to it, with a good announcing crew. The third is to read about it in the words of a good writer like Roger Angell, Jim Bouton, or Jonathan Schwartz. But the only way to understand a game of baseball is to run the numbers.

We can see all (or most) of the action of a game when we watch it in person. But, assuming we watch all 162 games of a season and get to know the players as well as we can, could we still spot the difference between those 0.250 and 0.300 batters that our mentor, Crash Davis, mentions? Would we recognize that "one extra flare a week" separates the star batter from the standard?

The dependence of understanding baseball (and by extension, rating or grading it) by the numbers is something that was recognized even before baseball as a sport went professional. Henry Chadwick, a writer who covered cricket for the *New York Times* in the 1850s, fell in love with baseball and adapted the summary used for cricket into the earliest regular box scores for his adopted sport ("Henry Chadwick"; Wikipedia, "Henry Chadwick [writer]"). By 1860, he was editing the first baseball annual, *Beadle's Dime Base-Ball Player* and including in it the basic statistics (totals of hits, bases, runs, and home runs) tallied by players on the more famous club teams of the day. Given the uncertain scheduling of games, players would have unequal numbers of opportunities to achieve hits and home runs, so Chadwick also began to calculate rate statistics such as runs per plate appearance, so that the high scorers could be identified on a standardized scale (Schwarz, 8). Chadwick also introduced the method for keeping score in a baseball game to assist in getting the statistics correct, including the use of "K" for strikeouts. (No one knows why he picked K, although there are guesses that it stands for the last letter of struck) ("Henry Chadwick [writer]").

When we talk of sabermetrics, we are talking of statistics, statistics for baseball, saber from the Society for American Baseball Research (SABR), and "metrics" from measuring tool. The statistics that we use can be highly complicated in a quest to uncover the value of a player to a team, or they can be very simple as in the basic counts (or frequencies) of a player's performance: How many times at bat? How many hits? How many outs? Even the most involved formulas for player evaluation come back to those simple totals, whose tabulation and verification can involve years of effort. Hack Wilson set the modern baseball record for runs batted in (RBIs) in 1930, with a total of 190 except that total was incorrect. SABR researchers found an uncredited RBI in the box scores, pushing Wilson's record total to 191; a claim that was accepted by Major League Baseball (MLB) in 1999 ("Hack Wilson"; Schott).

That process of collecting the observations, finding, and recording the raw numbers is a key one that many of us who depend on and delight in sabermetrics often overlook. But, right at the beginning, Henry Chadwick knew to check and double-check the numbers and to make it easier for the score-keepers to do their job and collect the numbers. Whatever we do to refine that raw information into pure sabermetrics gold starts with those observers and their recorded numbers. And that dependence continues even today with the most modern and technological of performance capture systems. The basic listing for position players in a standard reference like Baseball Reference (baseball-reference.com) is:

Name	G	PA	AB	R	H	2B	3B	HR	RBI	SB	CS	BB	SO	BA	OBP	SLG	OPS	TB
Mike Trout	134	600	470	110	137	27	2	45	104	11	2	110	120	0.291	0.438	0.645	1.083	303
Christian Yelich	130	580	489	100	161	29	3	44	97	30	2	80	118	0.329	0.429	0.671	1.100	328

Of the 18 columns of measurements, 13 are counted statistics not calculated. We are working with their plate appearances, at-bats, runs scored, hits (total), extra-base hits by type (doubles, triples, home runs), RBI, stolen bases (SB), caught stealing (CS), walks (bases on balls, so BB), and strikeouts (SO and not here, though!). Then, we reach the calculated statistics.

From their counted stats for the 2019 season, we can get a sense that these two players are somewhat similar, with roughly similar totals of plate appearances, at-bats, runs, RBIs, and strikeouts. The Yelich fellow seems a better hitter (161 hits vs. 137) and a lot faster (30 SB vs. 11), but not as patient (80 walks vs. 110). Given that both players are superstars in their prime (both age 27 in 2019), who are often considered most valuable player (MVP) candidates (Yelich finished second in the National League (NL) vote; Trout won his third MVP in the American League (AL)vote), their total statistics alone present us with a good case for their equivalent status.

But can we pick apart their differences or confirm their similarities even better? There are still those minor discrepancies in opportunities (at-bats, plate appearance) that seem to make the direct comparison of totals a bit unfair to Yelich. So, if we want to place Yelich as the "better" batter, can we measure his superiority more finely? Henry Chadwick ran into this same issue, in more stark terms, when dealing with totals for players whose schedules might be half as long as their "superstar" fellows. His initial solution was to divide the totals by games played, but even there we run into an imbalance because individual games may offer different numbers of opportunities; high scoring games bring more batters to the plate (Schwarz, 11). The solution on which he (at the suggestion of another statistics enthusiast, H. A. Dobson) and we settled is batting average BA: Total Hits/At-bats, to give a closer look at individual batting skill. Yelich's BA for 2019 is: $161/489 = 0.329$; Trout's BA is: $137/470 = 0.291$. For their leagues as a whole, in 2019, the NL had a 0.251 BA, and the AL hit 0.253. Crash Davis was right on the money!

Notice, however, that we choose at-bats for the divisor and not plate appearances. Why? The difference in the totals comes from the idea of batting skill being based on the ability to "contact" (hit) the ball. And not every batting opportunity winds up including the opportunity to hit the ball. Walks come first to mind; a walk is an opportunity that is spoiled by bad pitching. So, it seems unfair to include those opportunities when determining batting skill. (Though not always. In 1887, walks were included as hits because, "a walk is as good as a hit" according to an old baseball saying. BAs from 1887 are notoriously left out of the record books) (Forman, "A note"). Eventually, other "spoiled" opportunities (hit by pitch, sacrifice flies, sacrifice bunts, being awarded first base on catcher's interference) were also excluded, creating our modern statistic, the at-bat, from the more basic plate appearance. Yelich, with fewer plate appearances (580) has more at-bats (489), and so even though he has 17.5% more hits (161/137), his skill is only 13% higher than Trout's (0.329/0.291). As Crash Davis asserts, statistics like BA help us to "see" the game, but we, in turn, also need to see the statistic by learning to ignore parts of the game outside it.

A further refinement on the idea of batting skill is to again look at the idea of "contact." What do a player's contact opportunities tell us about his batting? The focus on contact was a primary interest in the 19th century, as baseball developed from other batting games like cricket or rounders. A truly skillful batter was one whose struck ball eluded the crowd of fielders. So, plate appearances without contact (walks, strikeouts, hit-by-pitch) or with too much contact (home runs, where the ball "eludes" the fielders by going out of play) do not appear to contribute to measuring the essential skill of "hitting 'em where they ain't" (to quote early 20th-century star Willie Keeler). Sacrifice flies (SF) get put back into play because they are balls that do not elude outfielders, even if they do score a runner. At-bats already eliminate walks and batters hit by a pitch (HBPs), so we simply continue whittling away the unwanted to get: Batting Average, Balls in Play (BABIP) = (Hits − HR)/ (AB − SO − HR + SF). For "pure" hitting, then, Yelich rates a BABIP of (161 − 44)/(489 − 118 − 44 + 3) = 117/330 = 0.355, while Trout comes in with a BABIP of (137 − 45)/(470 − 120 − 45 + 4) = 92/309 = 0.298. Yelich is clearly the 19th century "star," but how much of a star is he? The 2019 BABIP for both leagues was 0.298, so Yelich stands out and Trout suddenly looks average. What happened?

For one, Trout is being penalized more for home runs because his HR total (45) is a bigger part of his hitting (45/137 = 32.8%) than Yelich's (44/161 = 27.3%). So when we exclude the balls "out of play," Trout loses on the skill measurement. In addition, when we analyze what happens with a ball in play, we run into at least two other factors: the skill of the fielders and the chance for a batted ball to take a "bad hop" and elude the glove from luck rather than from skill. Not to say that fielding and luck are not part of the basic BA, but since we strip out some part of solo hitting (home runs) and some part of pitching skill (strikeouts), BABIP seems to put more of a premium on the part of hitting that leads to exciting plays. And the difference in BA and BABIP may give us insights into who is facing tougher fielding or is simply luckier. Over the past three seasons (2017, 2018, 2019), Yelich posted BAs of 0.282, 0.326, and 0.329, with BABIPs of 0.336, 0.373, and 0.355; Trout hit 0.306, 0.312, and 0.291, compared with BABIPs of 0.318, 0.346, and 0.298. From those numbers, we could say that 2019 was not Yelich's "lucky" year (+26 point difference, compared with his league's difference of +47) and Trout, like most power hitters, was downright unlucky (+7 point difference, against the AL's + 49). What we are measuring with BABIP, though, is excitement, putting the ball into play and standing on a base after the dust settles. Power hitters need not apply.

In fact, to this point, we have been ignoring power as a component of hitting. With BA and BABIP, we treat all hits as equal, one hit is one success for the statistic. But clearly, some hits are more equal than others, a fact that Henry Chadwick recognized early on as well (Schwarz, 11). He collected total bases as a measure of hitting total value, and we still do. The "weights" we assign are the obvious ones: each base gained by a hit is the value of the hit, so a single is worth 1, a double 2, a triple 3, and a home run 4 bases. Over the course of 2019, Yelich's hitting amounted to $(161 - 29 - 3 - 44) \times 1 + (29) \times 2 + (3) \times 3 + (44) \times 4 = 328$ total bases. Trout contributed $(137 - 27 - 2 - 45) \times 1 + (27) \times 2 + (2) \times 3 + (45) \times 4 = 303$ total bases. Again, we run up against the issue of one player having more opportunities and thus making totals a less obvious comparison of skill or value created, so we divide by at-bats to arrive at Slugging Percentage (SLG) = (Singles + Doubles $\times 2$ + Triples $\times 3$ + Home Runs $\times 4$)/At-bats. Yelich's SLG is $328/489 = 0.671$, and Trout slugged $303/470 = 0.645$, so Yelich wielded the heavier bat both in total bases and per at-bat.

Or did he? Power hitting as a topic focuses on extra bases, the bases achieved beyond the first, so to see it more clearly we want to consider only the extra bases made through doubles, triples, and home runs. To that end, we leverage the fact that BA treats all hits as equal (worth 1) and remove it from the SLG, creating a statistic called isolated power (ISO) = (SLG − BA). With ISO, we have the rate at which a batter creates extra bases per at-bat. For 2019, Yelich had an ISO of $(0.671 - 0.329) = 0.342$, while Trout generated an ISO of $(0.645 - 0.291) = 0.354$. Trout, we can see, brought a bit more "power" to the plate than Yelich.

Another issue about our measuring batting skill is that we have been ignoring patience, or "plate discipline." "Wait for your pitch!" we tell batters, thus encouraging them to draw walks if the pitcher is dancing outside the strike zone. But, batting average (BA) ignores walks, as does at-bats. One of the major changes in baseball's use of statistics was bringing walks back in. To be honest, they never really left but simply did not garner much attention until the Oakland A's under general manger (GM) Billy Beane in the 2000s recognized how undervalued they were and started stocking their roster with underpaid but above-average batters, as detailed in Michael Lewis's famous account of the underdog A's, *Moneyball* (2003).

The statistic that came into the limelight is On-Base Percentage (OBP) or On-Base Average (OBA, which confusingly is also used for pitchers as Opposing Batting Average), which brings contributions other than hits (Walks, Hit-by-Pitch) into the success part of the equation, although it also penalizes for productive outs (Sacrifice Flies): OBP = (Hits + BB + HBP)/ (At-bats + BB + HBP + SF). Roughly speaking, OBP will be 0.070 higher than BA for the "average" batter. More patient batters will have a difference higher than 70 points, whereas less patient ones will be less than 70 points. On rare occasion, BA will be higher than OBP, for a batter who is allergic to taking a walk.

How do Yelich and Trout fare when we look at their patience? Pretty darn good, in both cases. Yelich has an OBP of (161 + 80 + 8)/(489 + 80 + 8 + 3) = 249/580 = 0.429, while Trout reaches a bit higher with an OBP of (137 + 110 + 16)/(470 + 110 + 16 + 4) = 263/600 = 0.438. By comparison, the 2019 league OBPs were 0.323, putting Yelich (0.429 − 0.323) = 0.106 and Trout (0.438 − 0.323) = 0.115 points above average.

With OBP, we have a measure of how well a batter contributes by putting himself in position to score a run, and with SLG, we have a measure of how well a batter pushes runners along the way to score. It would seem like a good idea to combine them together and get a unified measure of batting excellence. The simplest version would be to add them together, creating OPS (On-base + Slugging). We trip over a number of issues, though. The first one is the difference in the scales of each component. OBP sits around an average of 0.323, but SLG is usually 100 points higher (in 2019, the NL SLG was 0.431, the AL SLG was 0.439). Furthermore, there is more value in a point of OBP for its effect on run scoring than a point of SLG (most analysis shows that a point of OBP generates 1.8 as many runs as an extra point of Slugging does) (Slowinski, "OPS and OPS+"). Adding them together "as is" treats them as equal and erases the higher value of OBP. And finally, if two players have identical OPS scores, then in the context of the statistic, they are effectively identical batters, even if one is all "excitement" and speed (Yelich: OPS = 0.429 + 0.671 = 1.100) and the other is patience and power (Trout: OPS = 0.438 + 0.645 = 1.083).

Nevertheless, we do use OPS, mainly due to its ability to track a player's ability to create runs. We will look at more advanced statistics to directly understand how players create runs, but when we look back at our more basic statistics, it turns out that OPS has a better chance to track (or correlate) with the more complex calculations. So, OPS it is. Or, a slightly "standardized" version known as OPS+. With some statistics, their novelty prevents most fans from understanding their value on sight. Is an OPS of 0.700 good or not? To alleviate the confusion, we can place the statistic on a new scale where 100 is the average value, and for "+" statistics, values above 100 will be above average. (And, correspondingly, for "−" statistics, values below 100 will be preferable. Consider ERA−.) For OPS+, we need to standardize both of its components and then add them: OPS + = (OBP/League OBP*+SLG/League SLG*−1) × 100, where the divisors (League OBP*, League SLG*) are adjusted for the ball parks of the individual batters (Baseball Reference, "Batting Stats Glossary"). For Yelich, living in a slight hitter's park (Park Factor [PF] 101 in 2019), his basic OPS of 1.100 is standardized to 179; Trout hits in a pitcher's park (PF 97), so his standardized OPS+ rises to 185. Both players are creating runs approximately 80% to 85% better than the average player.

The attempt to put a player's value (as opposed to measures of his skill) into a single statistic like OPS led to a number of "all in one" types of calculations. One of the most well-known is total average (TA), created in the 1970s by Tom Boswell, sportswriter for the Washington Post (Boswell, 1982, 137 − 144). TA is an attempt to capture the ratio of all bases accumulated by a batter to the number of outs he generates: TA = Accumulated Bases/Total Outs = (TB + HBP+BB + SB)/(AB − Hits + CS + Grounded into Double Play). For Yelich, his 2019 TA = (328 + 8 + 80 + 30)/(489 − 161 + 2+8) = 446/338 = 1.320; for Trout, we get 2019 TA = (303 + 16 + 110 + 11)/(470 − 137 + 2 + 5) = 440/340 = 1.294. By comparison, the NL TA in 2019 was 0.714, and the AL TA was 0.723. If we standardize the individual TA scores, Yelich is (1.320/0.714) × 100 = 185, and Trout is (1.294/0.723) × 100 = 179 (which reverses their OPS+ scores, but still delivers a similar judgement on their overall value). So, why do we not use TA?

The answer lies in one of Boswell's own statements: "Thus, the heart of TA: all bases are created equal." In sabermetrics, this idea has been tagged as the "Bases Fallacy" (Baseball Reference, "Bases Fallacy"). Why? Because all events that create bases do not create runs at the same rate. And runs (which lead to wins) are a better measure of a player's value than bases taken out of context.

The greatest advance in statistics in sabermetrics is the shift from bases to runs, and it is not surprising to see the major pioneers of sabermetrics all involved in some way with devising new runs-related statistics. In the 1970s, Bill James, the "father" of sabermetrics, began his work on runs created (RC), an "all in one" statistic that would bring together a batter's component totals and give an estimate of how many runs the batter produced for his team (Wikipedia, "Runs created"). The simplest formulation, RC = (A × B)/C, expresses Mr. James's straightforward approach (which most of his work embodies) that runs come from putting someone on base (A), pushing them along (B), and finding the rate at which they can produce runs given their opportunities (C). If this approach sounds a lot like taking OBP and SLG together, divided by Plate Appearances, well, it is. Except, in place of adding OBP and SLG, with RC we are multiplying the "putting on" and "pushing along" factors. The basic

RC formula is: RC = (Hits + BB) × (TB)/(At-bats + BB), which is used for any season before 1955. By 1955, baseball tracked more statistics for individuals (such as CS totals starting in 1951 and Intentional Walks [IBB] from 1955 and on), so the RC formula can be made a bit more "technical": RC = (Hits + BB − CS + HBP − GDP) × [TB + 0.26 × (BB − IBB + HBP) + 0.52 × (SH + SF + SB)]/(AB + BB + HBP + SH + SF). One aspect to notice is that the "technical" formula moves away from the Bases Fallacy by no longer assuming "all bases are created equal," with an (unintentional) walk here only worth 0.26 of a "hit" (single) and a stolen base only 0.52, at least towards "pushing along" a run. Mr. James derived these values from his work trying to fit team RC to the actual totals of runs scored in a season.

How well did Yelich create runs in 2019? According to the technical formula, RC = (161 + 80 − 2 + 8 − 8) × (328 + 0.26 × [80 − 16 + 8] + 0.52 × [0 + 3 + 30])/(489 + 80 + 8 + 0 + 3) = (239 × 363.88)/580 = 150.0.

For Trout, his RC value is: (137 + 110 − 2 + 16 − 5) × (303 + 0.26 [110 − 14 + 16] + 0.52 × 0 + 4 + 11])/(470 + 110 + 16 + 0 + 4) = (256 × 339.92)/600 = 145.0.

Now, how trustworthy are these estimates? If we compare with their teams in 2019, we have the Milwaukee Brewers with a technical RC value (for non-pitchers) of 819.5, against an actual total Runs of 757 (819.5/757 = 1.08, so we are "off" by 8%), and the Los Angeles Angels scored a 763.1 RC against 769 total Runs (763.1/769 = 0.99, off by 1%). The estimates land close to the actual totals, but there is enough "wiggle room" that we should not lean too heavily on Yelich's 150.0 beating out Trout's 145.0.

Other sabermetricians took Mr. James's approach further, looking at the run value of any of the standard events (hits, walks, steals, outs) as demonstrated by their game effects. In *The Hidden Game of Baseball* (1984), John Thorn (MLB's official historian) and Pete Palmer devised a method they called "Linear Weights" to assign values to singles, doubles, triples, and home runs, created from Mr. Palmer's simulations of games from 1901 to 1977, combined with an analysis of World Series games to capture the frequency of otherwise unrecorded events (e.g., the chance of a runner on first ending up on third after a single versus stopping at second) (Thorn and Palmer, 65). Since an out does not (usually) score a run, what value does it possess? Mr. Palmer set the value of an out to a negative score such that the overall sum of every event times their frequency (the "expected value" of the simulation) would be zero. The values did vary slightly over the different eras of baseball that Mr. Palmer simulated, but the overall simulation produced a single formula for batting runs that they used to estimate player contributions: Batting Runs = 0.46 × (Singles) + 0.80 × (Doubles) + 1.02 × (Triples) + 1.4 × (Home Runs) + 0.33 × (BB + HBP) + 0.3 × (SB) − 0.6

× (CS) − 0.25 (AB − Hits), where (AB − Hits) is the approximation for outs made.

According to Thorn and Palmer, how productive are Yelich and Trout? For Yelich, we have Batting Runs = 0.46 × (161 − 76) + 0.80 × (29) + 1.02 × (3) + 1.4 × (44) + 0.33 × (80 + 8) + 0.3 × (30) − 0.6 × (2) − 0.25 × (489 − 161) = 81.8; for Trout, we get Batting Runs = 0.46 × (137 − 74) + 0.80 × (27) + 1.02 × (2) + 1.4 × (45) + 0.33 × (110 + 16) + 0.3 × (11) − 0.6 × (2) − 0.25 × (470 − 137) = 76.1. Yelich had the better year by Thorn and Palmer.

The idea of Linear Weights, calculating the run value of the basic batting events has caught hold among a large number of sabermetricians, and their calculations tend to incorporate their own run values. Fangraphs, a leading website for player information, applies Linear Weights to OBP, creating their own version, Weighted On-Base Average (wOBA), to remove the Bases Fallacy drawback from OBP (Slowinski, "Linear Weights"). With wOBA, a weighted version of RC is also possible, wRC, that is adjusted by the corresponding MLB wOBA to give a statistic where zero is the average: wRC = [(wOBA − League wOBA)/wOBA "scale" + League Runs/PA] × Plate Appearances (Slowinski, "wRC and wRC+"). In 2019, Yelich had a wOBA of 0.442 (compared with his OBP of 0.429), and MLB as a whole had a 0.320 wOBA (which had a "scaling factor" of 1.157 to put wOBA at the same mid-point as OBP), with 0.126 Runs/Plate Appearance. So, Yelich would have a wRC of [(0.442 − 0.320)/1.157 + 0.126)] × (580) = 134.2, not quite as high as his technical RC score of 150.0. wRC can be standardized to put playing time and PFs on the same basis, giving us wRC+ . For 2019, Yelich's wRC+ was 174.

For Trout, 2019 saw him produce a wOBA of 0.436, so his wRC turns out to be: [(0.436 − 0.320)/1.157 + 0.126] × (600) = 135.8, which standardizes to wRC+ = 180. So, by wRC+, Yelich and Trout are about 75% to 80% extra when it came to their run production in 2019, compared with the average major leaguer.

The Fangraph approach to Linear Weights is based on the work of Tom Tango, another eminent sabermetrician and the creator of wRC, who looked at the way events generate runs from their context of how many outs and how many baserunners. Given three outs per inning and three bases, there are 3 × 8 = 24 different situations in which a batter can find himself (bases can be: none on, with one on first, with one on second, with one on third, with two on first and second, with two on first and third, with two on second and third, and with three on all bases). From these 24 situations, Tango and others create a "matrix" of Run Expectancies (RE), that is, the average number of runs that will (should?) score, based on the results of games from previous seasons. For the 2010–2015 seasons, Tango developed the following RE matrix (Slowinski, "Linear Weights"):

Runners	0 outs	1 out	2 outs
_ _ _ (None on)	0.481	0.254	0.098
x _ _	0.859	0.509	0.224
_ x _	1.100	0.664	0.319
x x _	1.437	0.884	0.429
_ _ x	1.350	0.950	0.353
x _ x	1.784	1.130	0.478
_ x x	1.964	1.376	0.580
x x x (Bases loaded)	2.292	1.541	0.752

For a batter, we total up the changes in RE "state" from each of his plate appearances and award him the value of the change, plus any runs actually scored. So, with 0 outs and a runner on first (0.859), if a batter hits a single that puts the runner on third, he changes the state to a value of 1.784 (0 outs, x _ x), and so contributes (1.784–0.859) = 0.925 "runs." If he had homered, then the state would be worth 0.481 (0 outs, _ _ _), so he contributed (0.481 − 0.859) + 2 = 1.622 runs. For the Linear Weights, we need to find the RE values for every hit during the season, then divide by the total number of that type of hit to find its RE value. For the 2015 season, Fangraph calculated the Linear Weights as: BB = 0.55, HBP = 0.57, Single = 0.70, Double = 1.00, Triple = 1.27, HR = 1.65, Out = −0.26. The process can be repeated each season to develop Linear Weights specific to that season.

The RE values per play can also be combined for the season as well, producing a statistic labeled RE24, accumulating all the plus and minus changes in the 24 "base" states that a player produces in a season, to develop a measure of how many runs produced (or deducted, if the total is negative), compared with a league average of 0 (Weinberg, "RE24"). For 2019, Yelich had an RE24 of 64.32 runs produced, and Trout managed to score an RE24 of 66.96. So, here, patience produces.

One important note to add to our overview, too, is to notice that Linear Weights, in either form, are impossible to calculate unless we have play-by-play information about entire seasons. Although MLB has long had professional, official scorers, the general run of sabermetricians has not had access to those primary records, usually forced to make do with newspaper box scores and written accounts. In 1984, Bill James (who else?) organized a volunteer group called Project Scoresheet to track all the on-field activity of every MLB game. Although the original organization fell apart due to disputes over ownership and sale of the scoresheet data, several commercial firms continue the practice and make their data available to subscribers and sabermetric websites such as Fangraphs. In addition, Project Scoresheet inspired another volunteer effort, Retrosheet, which since 1989 has been working to create play-by-play scoresheets for all major league games prior to 1984. At the moment, Retrosheet has worked back to the 1918 seasons for both regular and postseason games, along with less-complete reconstructions from box scores dating back to 1904 (Retrosheet.org, "Retrosheet").

One more set of items to add is the statistics we use when the batter transitions to baserunner. Here, we are curiously focused on self-achievement, so SB is a statistic that dates back to the Chadwick era, but we have not noted "taken bases" (moving further on the bases than forced to, such as first to third on a subsequent single) even now, when we do track a fuller account of on-field activities. Plus, failure was not an option at least until 1951 when CS became an official statistic. So, although we can compile lists of all-time base stealers by total (season, career), we are less familiar with all-time stolen base percentages (SB/[SB + CS]), or even the fairly obvious notion of "Net SB" (SB − CS), to put in a statistical penalty for the indiscriminate base thieves. Net SB, while it does correctly reward judgement, does fall to a worse version of the "Bases Fallacy" because it equates one base (SB) to one out (CS). A key innovation triggered by the focus on run creation was uncovering that a CS event is approximately twice as costly as a SB itself is worth (Linear Weights puts the benefit of SB at +0.30 and the cost of CS as −0.60) (Thorn and Palmer, 65–66). Just as our baseball forebears learned how to see and value a 0.300 BA, that "one gork" difference, we have learned to appreciate that base stealers need to be successful at least twice as often as they get caught in order to add to the fortunes of their teams.

Playing catch

Fielding is the most difficult part of baseball to be captured statistically, in part because it is also one of the parts of the game that has been so greatly advanced and perfected from its original form. The early professional games in the 19th century, played without gloves, permitted fielding a fly ball on one hop for a "bounce out." With the addition of leather gloves to protect the fielding hand, the ability to scoop up grounders or snag fly balls became a regular part of the game, making it harder to distinguish the better fielders from their average teammates. Still, the basics of fielding statistics follow from the same items that Henry Chadwick included in his box scores: put-outs and assists, but not errors (which miscues did not concern him as much as fielders reaching the ball; Schwarz, 10).

Name	Age	G	GS	CG	Inn	Ch	PO	A	E	DP	Fld%
Mike Trout	27	122	121	110	1051.2	303	294	5	4	2	0.987
Christian Yelich	27	124	124	110	1093	236	225	7	4	2	0.983

Even early on, though, the use of totals was not fully successful because the various fielding positions permit differing number of opportunities to record an out (putout) or assist in making an out (assist). Outfielders are less busy than first basemen. From the totals, baseball authorities in the new-born NL developed the idea of Fielding Percentage: Outs/Chances = (Putouts + Assists)/(Putouts + Assists + Errors), thus "standardizing" the various positions. The problem comes when those Fielding Percentages all creep up to towards perfect (1.000). For 2019, Yelich had a Fielding Percentage of (225 + 7)/(225 + 7 + 4) = 232/236 = 0.983, whereas Trout turned in a Fielding Percentage of (294 + 5)/(294 + 5 + 4) = 299/303 = 0.987. Not much difference, by percentages. What we are measuring is less their "pure" fielding ability as much as their ability to be "error-free" because the only difference in the factors is the Error term. And that hidden focus on Errors is what leads us to move further afield in trying to measure glove work.

One approach is to ignore errors entirely, partly because errors are judgement calls by the game official scorer, who must determine if a ball in play could have been turned into an out by a player "making ordinary effort" (as the Official Rules comment explains) (Leppard, 2019, 124). The more pressing issue is that errors are not charged for balls not reached, so slower fielders are protected from themselves, which should not add to the measuring of their fielding ability. The concept of Range Factor (RF), introduced in 1872 by Henry Chadwick and one of his colleagues, Mr. Reed of the Philadelphia Athletics ball club, and reintroduced by Bill James, gives us a better focus on the fielder's contribution by taking Outs (Putouts + Assists) and dividing by Games (Schwarz, 10). In 2019, Yelich ranged for 232/124 = 1.87, while Trout covered 299/121 = 2.47, which we may expect for a centerfielder to out-range a rightfielder. RF also penalizes players who do not play a complete game, which is unfair for good glovers who may be introduced as late-inning fielding replacements, so a refinement of RF is RF/9, which standardizes RF to a value for a complete game of nine innings: RF/9 = 9 × (Putouts + Assists)/Innings Played. The RF/9 score for Yelich is: 9 × (232)/1093 = 1.91, where Trout's RF/9 is: 9 × (299)/1051.67 = 2.56.

The drawback on RF is we are still comparing a player only to himself, finding his rate of fielding outs per game. This measure can be compared across the leagues, so we can say that in 2019, Yelich was below average by comparing his RF/9 of 1.91 to the NL right fielder RF/9 average (1.98), and Trout was above average (2.56 vs. league RF/9 of 2.49). But we have not made any adjustments for how many times our players were faced with fielding opportunities. If the 2019 Brewers were a team whose pitchers gave up a lot of ground-ball hits, then Yelich did not get the chances that other rightfielders had to boost their RF/9. So, an alternative statistic called Zone Rating looked at the opportunities offered to each fielding position (zone) and determined a fielding percentage for the player based on the actual number of opportunities given. One immediate issue is the need to identify those batted opportunities, so Zone Rating depends on having each batted ball assigned to a zone on the playing field. No surprise that this statistic was developed by John Dewan, the director of Project Scorebook and its subsequent for-profit operation (STATS, Inc.) (Smith, "What is Zone Rating?"). A further issue arose as plays were sometimes made "out of zone" by fielders who are fleet of foot. Should those out-of-zone plays (OOZ) be included in the zone fielding percentages? If so, then an outstanding play is being counted as a routine play, so fielders with great range were being underrated. The decision was to separate the two plays into different components: revised zone rating (RZR) and OOZ to separately rate a fielder on his reliability (RZR) or "surehandedness" and on his range (OOZ) (Slowinski, "RZR"). In 2019, Yelich had a total of 178 balls in zone (BIZ) hit at him, and he handled 162 of them, giving him an $RZR = 162/178 = 0.910$. He also ranged outside his zone to snag another 64 balls (OOZ). Trout in 2019 had 242 BIZ, turned 225 of them into outs, giving him an $RZR = 225/242 = 0.930$, with an added 69 outs made OOZ. NL rightfielders as a whole had an RZR of 0.905 and (with Yelich's innings played) would have made 76 plays OOZ, so Yelich had an average glove with below-average range. In the AL, centerfielders rated an RZR of 0.919 and would be expected to make 73 plays OOZ (based on Trout's innings played), making Trout a better-than-average glove with average range.

Even with the Revised version, though, we stumble a bit to get a measure of fielding for a few reasons. One is based on the arbitrary assignment of zones to fielders. Although based on years of conventional wisdom, the rise of repositioning (shifting) against batters with pronounced preferences for sides of the playing field (e.g., left-handed batters who pull exclusively to right) has undermined the conventional assignment of fielders with short-stops playing atop second, second basemen in short right field, and truly distorts the concept of "out of zone" plays. Another issue is the equivalence of batted balls (the Balls Fallacy?), because RZR considers any ball in the zone as an opportunity, no matter how hard or soft that ball is hit or however fast or slow it may be moving. In consequence, fielding measures themselves have shifted from the perspective of where the ball winds up to the prior condition of how the ball started on its journey there.

Ultimate zone rating (UZR) was created by Mitchel Lichtman to correct the simplicities of RZR and pick up several other parts of fielding (including throwing skill and double-play frequency) as well as incorporate error avoidance (Smith, "What is Zone Rating?"). It also uses a Linear Weights approach to translate the fielding activities into their equivalent "runs saved" values. It is scaled so that 0 represents the average fielder for a given position, and positive values are the runs that fielder prevented from scoring and negative the runs allowed that an average fielder would have prevented.

The key idea behind UZR is a credit/debit system for each type of batted ball, relying now on the video recorded for each play by MLB or one of the commercial sports recordkeeping companies (STATS Inc.; Sports Info Solutions, SIS) (Lichtman, "The FanGraphs UZR Primer"). Each batted ball is designated as one of four types (ground ball, bunted ground ball, outfield line drive, outfield fly) and assigned one of three speeds (slow/soft, medium, fast/hard) as well as a location on the field where the ball did or would land ("tracking by pixel") (Zimmerman and Basco, "Measuring Defense"). A total of six seasons' worth of batted balls is used to create the probabilities of a ball in each of these "buckets" (differentiated also by batter handedness, speed, and power) being caught by each possible fielder. Along with the catch probabilities, a batted ball also has a run value assigned to it (regardless of type of hit or baserunner situation to create a "fielding neutral" statistic). If a fielder records an out, he is awarded the value of the batted ball, calculated as its "hit" value minus its "out" value (which is negative, to set the average value to 0), times the probability of **not** catching it, to give the expected amount of runs his catch "saves." So, a ball hit to a particular location with a 30% chance of being caught, with a value of $0.50-(-0.25)$ or 0.75, would give a credit of $(1 - 0.30) \times (0.75)$ $= 0.7 \times (0.75) = 0.525$ to the successful fielder. What if it lands for a hit? Then, the debit would be shared by the flubbing fielders. So, if 10% of the 30% catches are by the centerfielder and 20% by the rightfielder, the centerfielder gets a debit of $(0.10) \times (0.75) = -0.075$ and the rightfielder takes $(0.20) \times (0.75) = -0.15$. The probabilities of catches are also adjusted by ball park to account for the differences in playing surfaces or outfield wall configurations or atmospheric conditions (altitude, humidity) that can affect fielding. The total of all such credits/debits over the course of a season produces the "Range Runs" (RngR) component of the UZR rating, which is combined with rating for throwing (ARM), double-play execution (DPR, used for infield positions), and error avoidance (ErrR). Top fielders will save approximately 15 runs per season for their teams.

For Yelich, his fielding credits/debits gave him with a RngR of 2.1, while he was about average with his throwing (ARM $= -0.1$) and in not making errors (ErrR $= -0.1$), for a total UZR of 2.1. Because he played only 1093 innings in the field, we can "standardize" his UZR for a more complete season of 150 "defensive" games (which is built on the average number of chances at the position given the player's innings, divided by the average number of chances at the position per game) to get his UZR/150 rating $= 150/(245/2.017) \times (2.1) = 2.6$ (which actually rounds to 2.5). For Trout, 2019 was a disappointing season in the field, as his

RngR came to −3.4 and he did not avoid errors well (ErrR = −0.7), but his throwing saved him somewhat (ARM = 3.2), giving him a slightly below average UZR of −0.9. Again, "standardizing" him for 150 games (from his base of 303 chances in CF), his UZR/150 = 150/(294.4/2.49) × (−0.9) = −1.14 (again, −1.2, due to rounding).

One caution we use with UZR is the amount of games needed to fill all those batted ball "buckets" is rather heavy (hence the use of six seasons for the baseline Linear Weights and catch probabilities). So, UZR may not be terribly accurate with less than a full season of fielding numbers, and it has a tendency to "swing" between seasons, particularly if the seasons are shortened. Mike Trout in 2018 (1076.67 innings played) rated 4.0 in UZR (5.6 in UZR/150) but dipped down to −4.4 (−6.7 UZR/150) in 2017, his shortest season (948 innings played) since becoming a starting outfielder. Max Kepler of the Minnesota Twins had the highest UZR for outfielders (minimum 1,000 innings played) in 2019 with 12.7 (15.9 UZR/150); in 2017, Kepler rated 2.2 UZR (2.1 UZR/150).

While Mr. Lichtman was developing UZR from the original zone rating, John Dewan (moving to a new company, Baseball Info Solutions, now Sports Info Solutions) took his own approach to a comprehensive fielding measure and came up with Defensive Runs Saved (DRS). Much of DRS is similar to UZR, in that both start from the perspective of the batted ball and how it is (typically) handled to award credits/debits (first called the "Plus/Minus" system in DRS, then the "Range and Positioning" system) to the fielders. What distinguishes DRS is a finer distinction of "buckets": balls are tracked along 260 different directions ("vectors") by speed (slow/medium/fast), distance traveled, and type (with DRS adding the infamous "fliner" for outfield line drives/"flat" fly balls) (Sports Info Solution, Inc. Fielding Bible, "Overview of the Plus/Minus System"). DRS is also less forgiving of flubs, because the "minus" for a missed chance is assigned entirely to the most likely suspect, the fielder with the highest catch probability for the ball. On the other mitt, DRS also recognizes positive fielding contributions that do not get captured in the standard totals, which it calls "Good Fielding Plays," based on the judgement of its play-by-play recorders, as well as broadening the concept of "error" into "Defensive Misplays" (e.g., throwing to the wrong base, which may not be awarded an error by the official scorer).

The more "customized" DRS ratings tend to run a bit higher than UZR, with the top fielders in the 15 to 20 DRS category. In 2019, Yelich was dinged by DRS with a rating of −2 for his fielding, with Trout suffering the same fate of having a −2 DRS rating. Again, as with UZR, we can see large "swings" from season to season, with Trout in 2018 having a +9 DRS but a −5 for 2017. UZR and DRS tend to agree in overall "direction" if not in the particular values that they assign to fielders. For 2019, DRS put Victor Robles of the Washington Nationals atop the OF (minimum 1000 innings) fielding leaderboard with a rating of +25 (7.0 UZR; 7.3 UZR/150). Max Kepler was 13th with a + 8 DRS score.

Even with the more finely graded batted ball measurements, DRS ran into the same problem that RZR has: shifts! Players being rated based on "position" is problematic when those positions are not where they used to be. Again, the advance of technology suggests a solution. MLB has been precisely recording the on-field action (and over-the-fence action for everyone not named "Judge") with the Statcast system that combines high-definition video with radar position finding. The result is a continual series of "snapshots" of the field showing positions, directions, and speeds of both ball and players. So, fielding skill can be determined not from more arbitrary zones or buckets but from actual starting/ending positions of ball and glove. In 2016, beginning with outfielders, MLB began tracking each fly ball and each fielder's starting location to develop a measure of catch probability for every ball hit to the outfield, based on distance needed to be covered (assuming correct tracking by the outfielder) and time allowed for the opportunity to catch the ball (beginning with pitcher's release and not time of bat contact, to approximate fielder adjustments based on their recognition of pitch types). With infield grounders, distance to the target base after fielding the ball and runner speed are factored in as well to assign the out probability. For informational purposes, MLB totals up 5 different categories of fielding chances based on their Catch Probabilities, from 5-Star (0% to 25%) through 4-Star (26% to 50%), 3-Star (51% to 75%), 2-Star (76% to 90%), and 1-Star (91% to 95%). Anything more than 95% catchable gets no star (Petriello, "Statcast introduces Catch Probability for 2017").

To create an overall fielding metric, MLB uses a credit/debit approach like UZR and DRS, by simply crediting fielders with a (1 − Catch Probability) award for each out made and with a (− Catch Probability) penalty for not making the out. So, a routine batted ball (90% catchable) generates either a+0.10 award or a −0.90 penalty depending on how the fielder handles it. Outs Above Average (OAA) is the cumulative total of a fielder's credits/debits during a season (Petriello, "A new way to measure MLB's best infield defenders"). With the tracking ability of Statcast, OAA can also be made directional, showing how many outs a player makes on balls to his left/right/in front/in back. And with the assignment of the run value of a batted ball (0.75 for infielders, 0.90 for outfielders who face a lot more extra-base hits), OAA can be translated into Runs Prevented (just as UZR is) (Tango, "Statcast Lab: Is there a different run value needed based on the infield slice?").

How do the new OAA/Runs Prevented scores turn out for Yelich and Trout? According to MLB, in 2019, Yelich was below-average in right field, with a −4 OAA and −4 Runs Prevented (Aaron Judge was the highest rated RF, with +8 OAA and +7 Runs Prevented). Yelich did have two 4-Star catches, at least. In center field, Trout also totaled a bit below average

(−2 OAA, −2 Runs Prevented), well behind Victor Robles, who led all fielders (out or in) with +23 OAA, +20 Runs Prevented ("Statcast Outs Above Average Leaderboard").

With the advance in pinpointing ball and player positions, more fielding measures are adapting to the zone-less approach. In 2020, John Dewan's company, SIS, announced that DRS would be redone to replace the "Range and Positioning" system (the updated "Plus/Minus") with a newer analytical technique based on player starting point instead of assigned "position" (again, defeated by the plethora of shifts on plays). The PART system (Positioning, Air balls, Range, Throwing) is aimed at infield play (where the shifts occur) and "Range and Positioning" will be retained for outfielder rating (for the moment). Essentially, under the previous DRS approach, plays involving shifts were not included in player fielding rating, due to the uncertainty of assigned batted balls to positions instead of to players. With PART, each ball is assigned to the players within reach based on their (known) starting positions (Reiff, "Defensive Runs Saved – 2020 Update"). The result is an increase (sometimes dramatic) in the DRS ratings. Among examples provided, Matt Chapman, the third baseman of the Oakland Athletics, went from a pre-2020 DRS of 18 to a PART DRS of 34, while Chicago Cubs super-utility player Javier Baez saw a rise from 15 to 26 in his DRS. The new age of fielding stats is upon us!

On the mound

The role of a pitcher was not considered terribly important in early baseball. The title "pitcher" is a clue; the player's job was to pitch (i.e., toss underhand) a ball so that the batter (or striker) could take a whack at it and put it into play, where the real skills of fielding and running could be demonstrated (Schwarz, 14). As a result, the early measures of pitching tend to focus on the team's achievement, like wins (and losses) or runs allowed, rather than on the individual achievement of the pitcher. As pitching skill (speed, delivery) was allowed to creep into the game, statistics that total up individual effort begin to appear (innings pitched, strikcouts, walks) as well as modification to "team" statistics, so runs allowed is accompanied by earned Runs Allowed (which are defined by the official rules as runs scored "without benefit of an error or passed ball"), permitting a fan to compare the skills of individual hurlers with less of a view of the team fielding behind him, although the original focus remained on the fielders (Schwarz; 9, 27). Again, the discrepancies in playing time led to the emphasis on rate statistics, so earned run average (ERA) became the premier pitching statistic, to standardized the number of earned runs per 9 innings: ERA = (Earned Runs)/(Innings Pitched) × 9. Similarly, other totals can be standardized over

9 innings (a complete game) for strikeouts (SO9), walks (BB9), hits allowed (H9), or home runs allowed (HR9), allowing comparisons between both regular and part-time pitchers and between specialized roles (starters garner three times as many innings as relievers, typically).

Name	W	L	W-L%	ERA	G	GS	GF	CG	SO	IP	H	R	ER	HR	BB	IBB	SO	HBP	BF
Max Scherzer	11	7	0.611	2.92	27	27	0	0	0	172.1	144	59	56	18	33	2	243	7	693
Justin Verlander	21	6	0.778	2.58	34	34	0	2	1	223	137	66	64	36	42	0	300	6	847

For 2019, Max Scherzer of the Washington Nationals pitched 172{1/3} innings and faced 693 batters. He was credited with 11 wins and debited for 7 losses, so his winning percentage (W−L%) is 11/(11 + 7) = 0.611. He (and his team) allowed 56 earned runs over his innings, so he (or they) generated runs at a rate of (56)/(172.33) × 9 = 2.92 per game (ERA). He totaled 243 strikeouts, which puts him at a rate of SO9 = (243)/(172.33) × 9 = 12.7, while giving up 33 walks at a rate of (33)/(172.33) × 9 = 1.72 and home runs at a rate of (18)/(172.33) × 9 = 0.9, less than 1 per game. Over in Houston, Justin Verlander of the Astros tossed 223 innings in 2019, got credit for 21 wins and debit for 6 losses, saw 64 earned runs cross the plate, and amassed 300 strikeouts while allowing 42 walks and 36 home runs. His rate statistics, for comparison, become: W−L% = 21/(21 + 6) = 0.778; ERA = (64)/(223) × 9 = 2.58; SO9 = (300)/(223) × 9 = 12.1; BB9 = (42)/(223) × 9 = 1.7; HR9 = (36)/(223) × 9 = 1.45. So, Verlander fell victim to the long ball approximately 50% more often than Scherzer. And who was the better strikeout artist? Verlander had the total (300), but Scherzer had the rate (12.7 vs. 12.1).

But we still run into a problem in comparing Scherzer with Verlander: the different league environments in which they pitch, along with the different home parks. Particularly since the introduction of the designated hitter in the AL in 1973, AL pitchers face a greater scoring environment than NL pitchers, meaning that comparing ERAs between leagues can be a bit misleading. So, we standardize ERA, similar to how we standardize OPS to get OPS+, putting ERA on a 100-point scale. But, because ERA improves as it gets smaller, if our standardized ERA is better the higher it is, to give us ERA+, we need to take the league ERA and divide by the individual pitcher's ERA, adjusted by their PF. In 2019, the NL had a league ERA of 4.38, so Scherzer has an ERA+ of (4.38/2.92) × 100 = 150, adjusted to 157 (due to Scherzer working in a hitter's park with a PF of 104). Verlander, hurling in the AL, where the league ERA was 4.60 in 2019, in a slightly unfavorable park (PF of 102), sees his ERA of 2.58 turn into an ERA+ of $(4.60/2.58) \times 100 = 178$, adjusted to 179. Because ERA traditionally is better as it goes lower, we can also create a standardized measure that moves in the opposite direction: ERA−, where values below 100 are preferable. All we need do is flip the base calculation to (Pitcher ERA/League ERA) $\times$ 100, which would give Scherzer an ERA− of 65 and Verlander an ERA− of 58.

Curiously, as batting statistics moved toward the "runs" perspective, pitching stats have moved in the opposite direction, away from runs and towards the more isolated components that pinpoint the pitcher's contribution. Strikeouts and walks help to focus on the individual pitching aspect of the game and away from the fielding ("team") elements, so the concept of strikeout to walkout ratio (SO/W) has been popular for measuring pitching prowess. Breaking down runs into their "basic parts," hits and walks, also gives another way to measure pitching skill as the rate of walks and hits allowed per innings pitched, WHIP = (walks + hits)/IP, where, as with ERA, we look for lower values as signs of better pitching. Of course, as Bill James noted, every run scored by the batters is a run given up by the pitcher and the fielders, so simply reversing batting statistics can give us insight on pitching performance, bringing us to measure opposing batting average (abbreviated as OBA, which can be easily be confused with on-base average, which is a strong reason to prefer OBP for the batting statistic) for the BA "given up" by a pitcher, as well as OBA of balls in play (BABIP, same acronym as for batters). Looking at Scherzer in 2019, he created a SO/W ratio of $243/33 = 7.4$ (compared with an NL league average of 2.7), had a WHIP of $(33 + 144)/172.33 = 1.03$ (NL WHIP = 1.32), had batters hit OBA $= (144)/(693 - 33 - 4 - 8 - 2) = 0.227$ (NL BA = 0.250), with a BABIP of 0.321 (NL BABIP = 0.297). Seriously above average, with the exception of BABIP. For Verlander, 2019 showed his skills as SO/W = $300/42 = 7.1$ (AL league SO/W = 2.7), a WHIP of $(42 + 137)/223 = 0.80$ (AL WHIP = 1.35), with opposing batters hitting $(137)/(847 - 42 - 6 - 0 - 0) = 0.171$ (AL BA = 0.255), and a BABIP of 0.218 (AL BABIP = 0.299). Outstanding, all around.

The anomaly of Scherzer's BABIP looking so below-average when the rest of his measures mark him as top tier is a signal that helped to propel the latest innovation in pitching statistics. While working through a plethora of performance measures like opposing BA and BABIP, sabermetrician Voros McCracken in 2001 made a surprising discovery: BABIP does not repeat well from season to season. Top pitchers such as Hall of Famers Greg Maddux and Pedro Martinez could be leading in BABIP in one season, then down on the bottom of the list in the next. Mr. McCracken's conclusion was that, at least on balls in play, the pitching skill component was swamped by the fielding and the "lucky bounce" elements that determine if a batted ball is a hit or an out (McCracken, "Pitching and Defense: How Much Control

Do Hurlers Have?"). Consequently, he and other sabermetricians began to turn to pitching events that removed fielding and luck from the calculation (as best we could determine), to create statistical measures that are known as "Defense Independent Pitching" (DIP) statistics. What the analyst relies on are the activities called (originally in jest) the three true outcomes: strike-outs, walks, and home runs.

The most widely used DIP statistic is named field-independent pitching (FIP), developed by Tom Tango and others, and it has the following expression: FIP = [13 × HR + 3 × (W + HBP) − 2 × SO]/IP + FIP constant, where the values of the coefficients (13, 3, 2) are the linear weights for each of the events (HR, W, SO) scaled so that balls in play equal 0 and then multiplied by 9 to give a result resembling ERA (Kincaid, "Evaluating Pitchers with FIP, Part 1"). Because "raw" FIP results tend to be centered on 0, unlike ERA, the final calculation is standardized to the ERA scale by adding the "FIP constant," which is equal to (League ERA − League "raw" FIP). For 2019, Scherzer had a "raw" FIP of: [(13 × 18) + 3 × (33 + 7) − 2 × (243)]/172.33 = −0.766, and the 2019 "FIP constant" was 3.214, so his final FIP was (−0.766) + 3.214 = 2.45. The NL as a whole had a FIP of 4.44 and an ERA of 4.39, by comparison. Verlander had a "raw" FIP of ([13 × 36] + 3 × [42 + 6]−2 × [300])/223 = + 0.054, giving him a final FIP of + 0.054 + 3.214 = 3.27 (vs. the AL FIP of 4.58 and ERA of 4.62). By "pure" pitching skill, FIP rates Scherzer better than Verlander.

Some further issues remain, though. Some sabermetricians object to using innings pitched in FIP, because an "inning" is based on three outs being made, and most of those outs are made by fielding (even as three true outcomes results continue to increase to more than one-third of plate appearances) (Patton, "MLB: Are the Three True Outcomes Killing Baseball?"). A stronger objection is to the use of Home Runs, because the shape of the park clearly has an effect on whether a fly ball is an out or a dinger. To correct for the park's participation, Dave Studeman replaced actual HRs with "expected" HRs (basically, the expected number of fly balls that a pitcher surrenders that would turn into a home run at the league "fly ball to home run" rate), giving us the "expected" FIP (xFIP): (13 × [Fly balls × League HR/FB%] + 3 × [W + HBP] − 2 × SO)/IP (Slowinski, "xFIP"). In 2019, Scherzer had a total of 155 fly balls, and the NL had a 15.3% rate of fly balls sailing over the fences, so his expected HR total would be 155 × (0.153) = 23.7, a bit higher than his actual (18), producing an xFIP value of ([13 × 23.7] + 3 × [33 + 7] − 2 × [243])/172.33 = −0.335 + 3.214 = 2.88, more than his "actual" FIP since he was aided by having fewer fly balls leave the park at his 11.6% actual rate. For Verlander, who also faced a 15.3% league rate in the AL, his number of expected HRs would be 225 FB × (0.153) = 34.4, slightly below his actual (36), making his xFIP = [(13 × 34.4) + 3 × [42 + 6]−2 × (300)]/223 = −0.038 + 3.214 = 3.18.

Now, how do we interpret what FIP and xFIP are telling us? Most saber-metricians will look at FIP and xFIP as better measures of "true" or under-lying pitching skill, absent the circumstances of actual playing conditions (fielding skill, park dimensions, lucky bounces) and so use them in compar-ison to ERA to find out how, at a given moment or across a given season, a pitcher is being treated by his circumstances. Scherzer in 2019 had an ERA of 2.92 and a FIP of 2.45 (xFIP = 2.88), so we may think of him as pitching "true" to his own skill level. Verlander, on the other glove, got an ERA of 2.58 for his time on the mound, but his FIP was 3.27 (xFIP = 3.18), signifi-cantly less impressive than what his circumstances produced. Any surprise to discover that the Astros rank highly as a fielding squad? (Rated number three in Team DRS in 2019 by the Fielding Bible) (Sports Info Solutions, Inc. Fielding Bible, "Team Defensive Runs Saved").

Take me out to the ball game

One recurring adjustment that we make is altering various statistical values by PFs, taking the effect of the park in which a player performs as a factor tied into the measurement of his performance. The debate on the need to measure park effects revolves around the allied, yet separate, concepts of skill and value. When we measure a player's performance without regard to his physical surroundings, we are measuring that player's value to his team. Imagine a left-handed power hitter getting 40 home runs for the Yankees, compared with a right-handed power hitter with 33 home runs. Clearly, the lefty has a higher value. But if we want to measure their skill as power hit-ters, we need to consider that Yankee Stadium "favors" lefties due to its famous right-field "porch" that stays more shallow than its left-field wall, offering about a 20% "bonus" to left-handed hitters when compared with righties (at least, in the pre-Aaron Judge days). If we adjust for that 20% bonus, our lefty is "really" producing 40/1.2 = 33 home runs, so their skill is the same.

For skill evaluations, then, we would like to account for the effect of the park. How? One method, used on the analytical website Fangraphs, tries to put the park and the team into a league-wide context, comparing runs scored at home with runs scored on the road (where the league, as a whole, is considered a "neutral" playing field, with no effect, plus or minus, on performance). So, the initial calculation is a ratio of: (Runs per Game at Home)/(Runs per Game on the Road + Runs per Game at Home). For the league context, with number of teams, any team will play (T−1) other parks at the "road" rate and one time at the (park) home rate, so the league value is (T−1) × Road+(1) × Home. The "park" context, to balance since a team plays equal numbers of games at home and on the road, must be (T × Home), giving us our "raw" PF of (T) ×

(Home)/[(T−1) × Road + Home] (Appelman, "Park Factors −5 Year Regressed"). Using past records (so that PFs are available for in-season updates), in the 5 previous seasons (2014–2018), the Milwaukee Brewers and their opponents scored 3555 runs at home in Miller Park in 408 games, so Home = 3555/408 = 8.713, and they and their opponents scored 3390 on the road in 402 games, so Road = 3390/402 = 8.433. Our "raw" PF is: (15 × 8.713)/[14 × (8.433) + 8.713] = 1.031. In Anaheim, from 2014 to 2018, the Angels and opponents scored 3379 runs at Home in 405 games, so Home = 3379/405 = 8.343, and they and their opponents scored 3666 on the road in 405 games, so Road = 3666/405 = 9.052, giving their park a "raw" PF of (15 × 8.343)/[14 × 9.052 + 8.343] = 0.927. At first glance, the Brewers have a "boost" at home of 3%, while the Angels "lose" about 7% from their stats at home. (Values greater than 1.0 are "hitter's parks"; less than 1.0 are "pitcher's parks." Note, too, that PFs are typically multiplied by 100 to eliminate the decimal point notation.)

Because the player performances are, generally speaking, half home and half road, the adjustment we should make should be similarly an average of the two. Because the road is neutral (i.e., equals 1.0), our adjustment would be (Home park + 1.0)/2, giving us an intermediate Park Factor (iPF). For the Brewers, we adjust their stats by (1.031 + 1)/2 = 1.0155; for the Angels, we have an intermediate adjustment of (0.927 + 1.0)/2 = 0.963.

Single-season stats are usually considered to be a bit volatile, so the final step is to account for the "weight" of our evidence by introducing a "regression" factor based on the number of past seasons we are using: 0.6 for 1 season, 0.7 for 2 seasons, 0.8 for 3, and 0.9 for 4 or more. (Appelman, "Park Factors −5 Year Regressed") In effect, we are giving our iPF a greater influence on the final value depending on how much evidence we have for its value being "true" (as opposed to the default 1.0 "neutral" value). Because we are using five seasons, our weight is 0.9 multiplied by the amount of PF adjustment that we have discovered: (1−iPF). For Miller Park, home of the Brewers, we get a final PF of: 1−(1 − 1.0155) × (0.9) = 1.014, or 101 when we remove the decimal point; for Angel Stadium, the final PF is: 1 − (1 − 0.963) × (0.9) = 0.967, or 97. Miller Park is a slight hitter's park (so Yelich's numbers will go down); Angel Stadium favors pitcher's a bit (so Trout's numbers will go up).

The process can be further extended to account for handedness, so we can calculate PFs for left-handed and right-handed batters separately, or it can be focused on certain kinds of events like home run hitting or strikeouts. Another change is to use the runs scored by opposing teams to see how the PF looks from the pitching perspective. Baseball Reference, another on-line repository of statistics and analysis, presents PFs for batting and for pitching separately (Baseball Reference, "Park Adjustments"). One issue we still need to consider, though, is the way in which performance is based on both time and place, not only on the park but also on the players (home and opposing) who are playing there. As we noted, before 2017, Yankee Stadium shows a bias for left-handed power. From 2017 on, though, right-handed power hitters hit 30% more home runs than lefties. What changed? The outfield walls? No. The outfielders, specifically, that guy in right field named Judge. Then again, Aaron Judge "broke" MLB's recording system, Statcast, so why shouldn't he also "bend" our sabermetrics as badly (Ruff, "Aaron Judge Seems to Break StatCast with Ridiculous Home Run")?

Rounding third and heading for home

With all of the sabermetric tools that have been developed, we have a very full tool kit with which to dissect the performances of batters, position players, and pitchers. But, we still leave one major question unanswered: Who's the best, overall? Which MLB star is the one for whom the era will be named in years to come? Putting together a measure of overall value (or skill) is one of the latest set of advances in sabermetrics that we can consider.

One of the first attempts was in 2002 by Bill James (naturally!), who took his normal, straight-forward approach and decided to figure out how to assign credit for team wins to each team member. Right away, we face that dilemma about how to rate "best" performance. Mr. James chose to look at value as the measuring stick: if a team wins more, then players on that team produce more value than players on teams that win less (even if their skill may be greater). The method Mr. James devised is called Win Shares, which is based on taking the number of wins, multiplying by 3, and then divvying up the Shares among each player based on his performance within the team context. The procedure for dividing up the Win Shares is founded on Mr. James's idea to look at marginal RC/saved by batters, fielders, and pitchers, and then to take a percentage of the total win shares based on each player's proportion of marginal runs contributed (Baseball Reference, "Win Shares"; Wikipedia, "Win Shares").

But what are marginal runs (MRs)? They are the runs produced above an expected low baseline for any team, approximately equal to {1/2} of the amount of runs expected to be scored or expected to be given up to the opposing team. In other measures, the baseline represents a certain bottom rung level of performance (replacement level is the term used). Mr. James does not specify what his baseline is meant to signify on an individual basis, only that it would be a team destined for a 0-win season (though his own work on runs converted to wins, the Pythagorean winning percentage method, which we discuss later, suggests his baseline is a 0.213 or 34, 35-win team) (Patriot, "Win Shares Walkthrough, pt. 6"). He calculates both baseline expected runs scored and runs allowed, then finds their differences from the actual Runs scored and allowed to get marginal scores, and proportions out the win shares by the percentage of the total each marginal run score represents, to give offensive win shares and defensive win shares.

Offensive win shares are further proportioned out for each batter, by finding the individual marginal runs for every batter, by looking at the baseline per batter (League Runs/Out × Outs Made × PF) × 0.52 (baseline). One further adjustment comes in with players who generate a negative marginal runs total, which would create a negative win shares score. As Mr. James does not find negative values of wins meaningful (as they seem to cross over into loss territory), he replaces any negative win share scores with 0, then proportions out the offensive win shares to each batter. The zeroing-out adjustment does lead to some unfortunate rescoring of win shares, though. Because the amount of marginal runs for the team increases with the removal of the negative scores, the calculation paradox that occurs is having lower win shares scores for the other, positive-scoring batters as their individual totals are now a smaller proportion of the team MR total. So, having poor-hitting teammates can hurt a player's value under win shares (Patriot, "win shares Walkthrough, pt. 3").

The defensive win shares are trickier to divvy up because they need to be proportioned between pitching and fielding first. Mr. James approaches the process through a series of "claim points," where each skill set puts in a claim on the total. To separate out pitching and fielding, there are 7 claim points needed, some that keep pitching and fielding together and some that are specifically based on one or the other. The first claim point is defensive efficiency record (DER), which is essentially a team's RF (outs/changes) and which combines both outs on batted balls (fielding) and "true outcome" outs like strikeouts (pitching). In consequence of covering both skills, this claim point is multiplied by 2 when added into the claim point total. Strikeout rate, walk rate, and home run rate cover the next three claim points (all pitching), before fielding moves in with error rate and "marginal" double plays (GIDPs above expectation). The last claim point is a "ballast" score (405 × team W−L%) to balance the contributions of pitching and fielding. The pitching% of the defensive win shares is the proportion of the pitching claim points out of the total claim-point amount, and the fielding% = (1 − pitching%). With a team that has the league averages for its claim point values, the result is pitching% is 67.5%, which tends to fit the general understanding of pitching being two-thirds of run prevention.

For individual pitchers, their pitching claim points are totaled and then the proportion for the individual compared with the team's total determines how many of the pitching win shares are handed over. What are the pitching claim points? Marginal runs allowed, individual won/loss/save scores calculated as (3 × Wins − Losses + Saves)/3), credit for "save equivalents"

(3 × Saves + Holds), and their batting MR (not zeroed out if negative). But if the individual claim point total is negative, then it is, itself, zeroed out (Patriot, "Win Shares Walkthrough, pt. 4").

Fielding win shares is the most complicated, involving differing claim point calculations depending on individual position, which claim point totals (scaled to 100 points) are further weighted by the run "cost" of each position (Catcher = 38; 1B = 12; 2B = 32; SS = 36; 3B = 24; OF = 58). As an example, for outfielders, the claim points include putout rate (40-point scale), the team's DER (again, on a 30-point scale), "arm elements" (assists and double plays above the league rate, scaled out of 20 points), and error rate (out of 10 points) (Patriot, "Win Shares Walkthrough, pt. 6").

Win shares is a balanced measure that relies on a good deal of Mr. James's baseball knowledge and intuition, which he openly published and discusses. That transparency may be its biggest drawback, alas, because it permits us to see the way in which the win shares gears turn, using scores and point scales that strike many sabermetricians as "ad hoc" judgement calls. The outcomes are almost always dependable (for 2019, Yelich and Trout both rated 33 win shares, tying them for second in MLB) and occasionally surprising (the person tying Yelich and Trout for second was Yankee super-utility infielder DJ LeMahieu; the person keeping them in second was Oakland SS Marcus Semien) (Reiff, "A Surprising Win Shares Leader"). But that openness and the lack of a general theoretical grounding (like deriving run costs from Linear Weights) just keeps moving analysts and fans away from win shares and towards other, more opaque measuring tools.

What did gain the attention of analysts and fans was wins above replacement (WAR) level, a similar all-in-one measure that attempts to measure how many wins a player contributes to his team beyond the baseline level of a "replacement" player (someone whose overall skills would not qualify for regular MLB play, but just below, at the level called "Quad A," "AAAA," to show that they are at the top of the minor leagues). WAR attempts to provide the same type of rating that win shares does, but it approaches the objective from a bottom up focus on RC (or runs prevented, for pitchers) by players as opposed to win shares working "top down" from wins to player contributions. WAR is also a "community" creation, being developed by several different sabermetricians on multiple discussion boards and forums, unlike the singular development of win shares. Unfortunately, the WAR community has not formed a single consensus, so the current state of the measure finds us with a choice of WAR ratings (bWAR, fWAR, WARP), offered by the major reference and analysis organizations and publications (Sports Info Solutions, Inc., FanGraphs, and Baseball Prospectus).

To understand WAR, we can look at one variety, bWAR from SIS on their Baseball Reference website, and compare it with its "colleagues." As presented on Baseball Reference, bWAR is calculated separately for position players and pitchers. For position players, there are six components needed, four of which are from a player's active performance: runs from batting, baserunning, double play avoidance (or not), fielding. To those are added a position credit/debit given for the amount of skill required to field each position, and an adjustment of "replacement" runs. The last component comes from having the previous components all based on "performance above average," where a zero-value would mean an average performance. So, to be "above replacement" requires a separate adjustment to rescale the measurements. Why not just leave it alone? Why not, indeed, so we also will see players rated in Wins Above Average (WAA), which is simply (WAR–"replacement" adjustment) (Baseball Reference, "WAR Explained").

For bWAR, the measure of runs produced comes from their weighted on-base percentage (wOBA) ratings, converted from bases to runs as weighted runs above average (wRAA). One additional category is added, as well, reached on error, which is considered an out for BA purposes but which does put a runner on and can contribute to scoring. With baserunning, both stealing/CS and advances beyond expected (going from first to third on a single, scoring from second on a single, etc., versus being thrown out or not going) are considered. The double-play component looks at the number of GIDP opportunities faced minus the league GIDP rate and minus the actual number of double plays into which the player hit. The fielding component was a zone rating (Sean Smith's Total Zone Rating) up to 2012, and it is still used for bWAR ratings up to 2003. Since then, though, bWAR has been based on SIS's DRS, which we previously discussed.

After these "active" components are calculated, a player's bWAR is adjusted for the runs value of his various positions, proportional to his defensive innings played. The runs values are estimated by taking the runs scored by each position across the league and assumes that the differences from position to position, plus or minus, are a reflection of the runs produced that are being given up in exchange for the runs saved by the fielders, in equal measure. So, catchers, with the lowest average run production, are assumed to represent a fielding credit of +9 runs, and shortstops see a +7 credit, while first basemen take a −9.5 debit to balance out their high run production. DHs get nicked for −15 runs because they have no fielding value. For NL pitchers who bat, they are given a position adjustment which is calculated to give pitchers overall a wRAA value of zero because pitcher hitting is considered to be an optional part of their game that neither contributes to nor harms their overall performance rating.

With his performance in 2019, Yelich would earn the following Run Components – batting: 56 runs above average; baserunning: 7; GIDP avoidance: 1; fielding: −2; position: −5. His total runs above average is:

$(56 + 7 + 1 - 2 - 5) = 57$. With 130 games played, he averaged $57/130 = 0.438$ RAA/9, with $64/130 = 0.492$ on the scoring side and $(-7)/130 = -0.054$ on the fielding side.

For pitchers, the active component is runs allowed, which is compared with an "expected" runs allowed (assuming an average pitcher in the same games as the pitcher to be rated), with adjustments for team fielding, PFs, and pitcher's primary role (starter vs. reliever), which takes into account that relievers have lower rates of runs allowed and so tend to skew the league average too low. To handle interleague (DH/non-DH) games, an adjustment of ± 0.2 runs per game is used if the DH is added (or removed) from the lineup.

What we have been adding up, though, are runs above (or below) average. How do we get from runs to wins? Here is one of the controversies with WAR, all varieties. Can we discover a way to take runs and meaningfully convert them into expected wins? We think we can, and the idea comes originally from an observation by Bill James (as does so much in sabermetrics). Mr. James uncovered a relationship in runs scored and runs allowed that he referred to as the Pythagorean winning percentage: Expected W−L% = (Runs scored)^2/[(Runs scored)^2+(Runs Allowed)^2]. Although not a geometric model, the use of squaring 3 times reminded Mr. James of the Pythagorean formula: $a^2 + b^2 = c^2$, which led to its nickname (Davenport and Woolner, "Revisiting the Pythagorean Theorem"). Further work by Mr. James showed that a better choice of exponent was 1.82, and a more dynamic version of the formula by a sabermetrician using the Internet ID of Patriot gives us an exponent of $x = $ (Runs/Game)$^{(0.285)}$ (Baseball Reference, "WAR Explained, Converting Runs to Wins"). The key connection, for bWAR, is that we can now "add" our player to be rated into an "average" (0.500) team, where we assume that their Runs scored and Runs Allowed totals per game are equal. In 2019, the NL had an average of 4.78 Runs scored per game, and Yelich added 0.438 runs above average, so an average NL team with Yelich added would see scoring of $(2 \times 4.78 + 0.438) = 10.0$ total runs per game, giving us an exponent of $(10.0)^{0.285} = 1.927$. Their Expected W−L% would be (when we separate out Yelich's batting and fielding contributions to runs scored and runs allowed): $([4.78 + 0.492]^{1.927})/([4.78 + 0.492]^{1.927} + [4.78 - (-0.054)]^{1.927}) = (24.62)/(24.62 + 20.83) = 0.542$, which is 0.042 better than average (0.500). So, over Yelich's 130 games, his "team" should see $130 \times (0.042) = 5.4$ WAA. (The actual calculation comes to 5.3, due to some rounding that we skipped over.)

But what about replacement level? To find Yelich's WAR, we need to add a credit for the difference between an average team and a replacement level team. The definition of replacement level varies by flavor of WAR used. For bWAR, the current definition is a team that would produce a W−L% of 0.294, because compared with such a team, the average MLB squad would produce (0.500

− 0.294) × 162 = 33.37 Wins, so MLB as a whole would produce (33.37) × 30 = 1001 (rounded to 1000) WAR (Baseball Reference, "WAR Explained"). To find a player's WAR credit, then, we simply proportion his replacement wins by his playing time (plate appearance or innings pitched). But all player roles are not created equal, nor are all leagues. So, the 1000 replacement wins are divided into a position player subtotal (59%) and a pitcher subtotal (41%), which split is apportioned based on the dollar amounts of all free-agent contracts signed in the previous four seasons. And league values are estimated based on records from interleague play and the performance changes of players who change leagues. Recent AL/NL differences have put the league values at 525 (AL) and 475 (NL) (Baseball Reference, "Position Player WAR Calculations and Details"). So, as a position player, Yelich is entitled to his share of 0.59 × (475) = 280.25 replacement wins. With 580 PA, he earns (280.25)/(93238) × 580 = 1.7 replacement wins, giving him a final bWAR of: 5.3 + 1.7 = 7.0. Over in the AL, Mike Trout earned a bWAR of 8.2, and our pitchers Scherzer (NL) and Verlander (AL) were given ratings of 5.7 and 7.4, respectively. Approximately, a bWAR of 2.0 is a regular, 5.0 is an All-Star, and 8.0 starts us talking about MVPs.

What happens when we switch to a different brand of WAR? Sometimes, very little. The differences are buried inside the particular details, essentially from the use of different measures to calculate the batting, fielding, and pitching runs, or just different versions of the same or related measures. The batting metric in bWAR is wRAA, and so it is in Fangraph's fWAR as well, whereas Baseball Prospectus goes straight back to Linear Weights to calculate the RC by the batter in their WARP statistic. For fielding, bWAR uses DRS, whereas fWAR relies on UZR, and WARP goes off on their own with fielding runs above average (FRAA), which uses a play-by-play, locational perspective similar to MLB's OAA (as opposed to a zone approach). For pitching runs, bWAR uses Runs Allowed, as does WARP, whereas fWAR goes after pitching with FIP. To convert from runs to wins, fWAR uses a simplified formula developed by Tom Tango related to Patriot's Pythagorean formula, based on the entire MLB run scoring environment and lacking any exponents. Another key, if minor, difference is the setting of replacement level, which for bWAR and fWAR is a 0.294 W−L% but it still set at 0.320 for WARP (a level which bWAR was previously set on) (Baseball Reference, "WAR Comparison Chart").

What do the differences amount to, for our selected players? According to fWAR, in 2019, Yelich was rated in runs above average as−batting: 56.7; baserunning: 8.5; fielding: 1.9; position: −5.8, which total up to 61.3 (vs. 57 for bWAR). He gains another 18.2 for replacement runs (18 in bWAR), and 0.5 to adjust the overall league to an average of zero, which puts his runs above replacement at 80.0. Converting to wins, his fWAR score is 7.8 (vs. 7.0 bWAR). Trout rated 8.6 fWAR (vs. 8.2), and our pitchers Scherzer and

Verlander landed at 6.5 (vs. 5.7) and 6.4 (vs. 7.4). If we look at WARP, we see Yelich is given 8.2; Trout has 8.9; Scherzer is awarded 6.2; and Verlander snags a 7.9. For consistency, we can see that all three measures like Mike Trout the best, but Justin Verlander could be either second best (by bWAR) or fourth (in fWAR).

Player	bWAR	fWAR	WARP
Christian Yelich	7.0	7.8	8.2
Mike Trout	8.2	8.6	8.9
Max Scherzer	5.7	6.5	6.2
Justin Verlander	7.4	6.4	7.9

WAR, Wins above replacement bWAR - *Baseball Reference Wins above replacement;* fWAR - *Fangraphs Wins above replacement;* WARP - *Baseball Prospectus Wins above replacement.*

In addition to the variation found among the differing flavors of WAR, the lack of transparency in how the numbers are calculated has also been a source of criticism for this type of measure. Even the need of WAR for the most advanced data collection methods, such as bWAR's fielding component relying on DRS, which requires having every batted ball assigned to one of 260 vectors, speed, distance, and landing location, puts the input requirements considerably out of reach for the typical sabermetrician, never mind the typical fan. The prevalence of WAR in discussion of player contract values and seasonal awards and career recognition by the Hall of Fame is a testament to how the concept has taken hold. But, the proprietary nature of both the component formulas and the general inaccessibility of the data has cast a shadow over how tightly the judgements that WAR presents can be embraced by the sabermetric and general baseball community.

Still, it is possible to create simplified versions of WAR that work from the readily available statistical totals found on every analysis site. Calculators abound on the Internet, along with detailed journals about attempts to work through the analytical verbiage to reproduce the "magic" formulas. One notable attempt was mentored by one of the contributors to this book for a senior undergraduate thesis that produced a version dubbed "SHU-WAR" (for Seton Hall University). SHU-WAR worked directly from the Linear Weights analysis of Palmer and Thorn, so that its batting runs above average were the individual linear weights RC minus the league average LWTS RC. Likewise, its baserunning, fielding, and pitching components were adapted directly from Palmer and Thorn's pioneering work. Positional adjustments were assigned from the most common values found among the current WAR varieties. And replacement value was pinned down easily by taking (plate appearances)/30 for position players and (innings pitched)/50 for pitchers, which generally results in getting an average of 20 replacement runs for a regular, an average used in the more sophisticated versions. Runs are converted to wins by using Mr. James's original Pythagorean winning percentage formula, with its exponent of 2, so that runs per win can be estimated by the total number of runs scored per game (which is 2 × League R/G). How well does SHU-WAR work? In its analysis of the performance of 234 players from different eras from the 1927 Yankees to the 2018 Dodgers, SHU-WAR's values were within 2 points of bWAR for all but 13 players. By comparison, bWAR and fWAR had 9 disagreements of more than 2 points among those players. In more technical terms, the linear correlation between SHU-WAR and bWAR was 94% (bWAR and fWAR had a 95% correlation score). In consequence, if we use SHU-WAR, we can "cover" about 88% of the change in bWAR values between players by the changes in their SHU-WAR values. (The rest of the change is lost to us due to factors that SHU-WAR is not capturing.) Not a bad result for an undergraduate (Hendela, "Sabermetric Analysis: Wins-Above-Replacement")!

Sidebar 1: the most valuable statistic?

Possibly the biggest break for the WAR statistic came in 2012 with the AL MVP award voting. Two clear candidates were present. For the old-time fans, the choice was Miguel Cabrera, third baseman for the pennant-winning Detroit Tigers, who completed a batting Triple Crown (BA: 0.330, 44 HRs, 139 RBIs), the first one in MLB since 1967. What's not to love?

Plenty, if you asked new-fangled sabermetric folks, who fell in behind the AL Rookie of the Year, Mike Trout, centerfielder for the Los Angeles Angels. Trout did not match Cabrera in the traditional batting categories (partly for not starting regular play until the end of April, limiting him to 139 games to Cabrera's 161), but he was not terribly far behind (second in batting with 0.326; 30 HRs; 83 RBIs), either. And in a number of other traditional statistics, he had the lead: first in Runs scored (129) and first in SB (49). Plus, he was acknowledged as the better fielder, playing a position as critical as Cabrera's (bWAR puts a +2.5 position adjustment on CF and +2 runs for 3B). Why not Trout?

The argument, split so neatly by traditional statistics, moved to the less-known older ones as well as the newer ones: OBP (Trout: 0.399, Cabrera: 0.393), slugging (Cabrera: 0.666, Trout: 0.564), on base + slugging (Cabrera: 0.999 for first in the league, Trout: 0.963 for second). But that only compounded the problem by adding more separate pieces to the puzzle. What people began to crave was a single answer to the question, Who's better? And that single-answer statistic looked to be WAR.

What did WAR say? It said "Trout," loudly, giving him a rating of 10 or better, whereas Cabrera came in with ratings at or slightly above 7. Although we readily admit that WAR's accuracy is not down to the tenths position, a difference of three points is significant. According to WAR, in 2012, Mike Trout was the best player in the AL, while Miguel Cabrera was bunched up with other worthy second-place finishers such as Robinson Cano, Adrian Beltre, and Justin Verlander. Unfortunately, WAR in 2012 was even less understood than it is now, so it was simple for old-school MVP voters to look right past it, alongside all the other confusing acronyms that were being tossed at them (adjusted OPS+, Trout: 168; Cabrera: 164). Cabrera prevailed, taking 22 out of 28 votes for MVP (Trout did tally the other 6), and finishing with 362 points to Trout's 281 (ahead of Adrian Beltre, with 210). Cabrera would repeat his MVP win in 2013, ahead of Trout again, although trailing in WAR again (only by 1.5 points, though, 7.5 vs. 8.9, according to bWAR). But since 2014, the WAR leader or runner-up has won the MVP (for either league) 8 times out of 12. WAR, what is it good for? Maybe choosing an MVP.

Did we make a mistake in 2012? We have to go back to the question, who is the most "valuable" player? Does WAR answer that question? As we noted with statistics like win shares, there is a distinction we make with our statistics between "skill" and "value." Many of the traditional statistics are value measures because they record what actually happened, which, as we said, is some mix of skill, context, and luck. Statistics that try to go beneath that surface of actuality, looking for expected results, are seeking to measure skill without the other two distractions. Neither approach is incorrect, but they are different, and we need to respect those differences and appreciate them. One lottery ticket may have been skillfully chosen with numbers that will likely produce a bigger payout should they hit. If it does not hit, though, it is just a piece of paper, while the ticket with the lucky numbers that do hit is the one with the better value, skill or no.

Oh, and win shares, which does try to measure value, liked Mike Trout (38) in 2012 a bit more than Miggy Cabrera (32).

Sidebar 2: stats and steroids, taken together

Can we use our statistical tools to discover if performance-enhancing drugs (PEDs), most commonly termed "steroids," are being used by the players (and, if so, which ones)? Until we know an authoritative "signal" that PEDs produce, then, no, we cannot definitively pin down the use of "juice." But, we can at least raise some questions as to the possible introduction of PEDs during a player's career.

One tool that has been applied to the question of PED use is the "aging" curve, which is perhaps best known in sabermetrics thanks to attempts by Bill James (yes, him, again) to determine the peak age of ballplayers (thought to be 28–32 before Mr. James attacked the question; long deemed to be 27 after his initial studies; still open to question due to the problem of aging players being almost always better-than-average, thus inflating the upper tail of the aging curve). One study of home run hitting that used the aging curve approach, "Cumulative Home Run Frequency and the Recent Home Run Explosion," appeared in the SABR annual (now semi-annual) publication, *The Baseball Research Journal*, in 2005. It was co-authored by three of the contributors to this volume, and it took the approach of looking at the cumulative HR rate (CHR, the accumulated percentage chance of a home run per at-bat) over the lifetime of those players who tagged 500 or more "dingers." They use as their touchstone Babe Ruth, the first great home run hitter, and they remark that Ruth's career shows a slow growth in his home run rate (from 6.5% in his early years, up to age 25, 7.9% by age 30, rising to 8.6% by age 33; and leveling down to 8.5% by the end of his career at age 40). Ruth is then compared with his successor at the top of the HR list, Hank Aaron who also demonstrated a slow growth HR curve over his career (5.0% up to age 25; 5.6% by age 30; 5.8% by age 33; and continuing to grow to 6.3% at age 39, declining to 6.1% at the end of his career at age 42). Both of these "sultans of swat" are a bit unusual because the majority of premium HR hitters show an HR rate peak at their prime (27–28 years old) and a subsequent decline. The strongest contrast, though, comes with the "new" kings of the HR derby (circa 2005), players such as Mark McGwire, where the HR curve not only continues to grow but grows substantially at the tail-end of his career (6.1% by age 25; 7.1% by age 30; 9.2% at age 35, finishing at 9.4% at the end of his career at age 37). What happened?

In the case of McGwire, we do have his (soft) confession to using PEDs as a way to recover quicker from his (admittedly frequent) injuries (which is the allegation used against pitchers who may have indulged in PED "self-medication"). So, perhaps we have uncovered a signal that steroid use is occurring, at least for home run hitters. But we run up against a problem that the signal is neither obvious nor consistent. Ruth and Aaron both show that growth in CHR alone is not enough; it needs to be a rather spectacular growth to separate out McGwire. And, worse, some PED cases do not show any growth at all. A follow-up study, "Cumulative Home Run Ratio and Today's Home Run Hitters," by Connor Love (Love 2012), a West Point student of two of the previous study's co-authors (Fr. Gabriel Costa and Col. Mike Huber) was published by SABR online in January 2012. Among other cases it examined was Manny Ramirez, whose CHR aging curve looked to the author to be similar to the CHR curve of Ted Williams, a "natural hitter" in the quoted opinion of Pulitzer Prize–winning sportswriter Arthur Daley. And Manny's curve does "flatten" nicely (5.6% up to age 25; 7.0% at age 30; 6.9% at age 35; finishing up at 6.7% at the end of his career at age 39). So, is Manny a natural hitter? He retired at 39 while facing his second suspension for testing positive for PEDs, which would require him to sit out 100 games. He has admitted to using PEDs. Unfortunately, our statistical tool, CHR, does not point him out as a possible user. But failure can be a recurring oddity with a statistical tool. Just like "Manny being Manny," sometimes "the numbers" are just going to be "the numbers," and we have to understand when they are telling a tale that does not really match the reality we are trying to pin down.

References

Appelman, David. "Park Factors – 5 Year Regressed". FanGraphs Library, library. fangraphs.com/park-factors-5-year-regressed. Accessed on November 15, 2020.

"Bases Fallacy", Baseball Reference, www.baseball-reference.com/bullpen/Bases_ Fallacy. Accessed November 2, 2020.

"Batting Stats Glossary", Baseball Reference, www.baseball-reference.com/about/ bat_glossary.shtml. Accessed September 20, 2020.

Boswell, Thomas. "Welcome to the World of Total Average, Where a Walk is as Good as a Hit", in *How Life Imitates the World Series*. Penguin Books, 1982.

Bull Durham. Written and directed by Ron Shelton, performances by Kevin Costner, Susan Sarandon, and Tim Robbins, Paramount Pictures, 1988. Quoted on Filmsite Movie Review, "Bull Durham (1988)", https://www.filmsite.org/bulldurham3. html. Accessed June 7, 2021.

"Chadwick, Henry", National Baseball Hall of Fame, baseballhall.org/hall-of-famers/chadwich-henry. Accessed September 20, 2020.

Costa, Fr. Gabriel, Michael Huber, and John T. Saccoman. "Cumulative Home Run Frequency and the Recent Home Run Explosion." Baseball Research Journal, 2005, pp. 37 – 41.

Davenport, Clay and Keith Woolner. "Revisiting the Pythagorean Theorem: Putting Bill James's Pythagorean Theorem to the Test". Baseball Prospectus, www.baseball-prospectus.com/news/article/342/revisiting-the-pythagorean-theorem-putting-bill-james-pythagorean-theorem-to-the-test/. Accessed on November 15, 2020.

Forman, Sean. "A note about statistics from 1887 and 1876." Baseball Reference, www.baseball-reference.com/about/note76-87.shtml. Accessed on September 20, 2020.

"Hack Wilson", Wikipedia, en.wikipedia.org/wiki/Hack_Wilson. Accessed September 20, 2020.

Hendela, Karl. "Sabermetric Analysis: Wins-Above-Replacement". Unpublished senior thesis, Seton Hall University, Department of Mathematics/Computer Science (Prof. John T. Saccoman), May 11, 2019.

"Henry Chadwick (writer)", Wikipedia, en.wikipedia.org/wiki/Henry_Chadwick_ (writer). Accessed on September 20, 2020.

Kincaid. "Evaluating Pitchers with FIP, Part 1." 3-D Baseball, www.3-dbaseball. net/2009/10/evaluating-pitchers-with-fip-part-i.html. Accessed on November 8, 2020.

Leppard, Tom, editor. Official Baseball Rules, 2019 edition. Office of the Commissioner of Baseball, 2019. Downloaded from https://content.mlb.com/ documents/2/2/4/305750224/2019_Official_Baseball_Rules_FINAL_.pdf on May 27, 2020.

Lewis, Michael. Moneyball: The Art of Winning an Unfair Game. W.W. Norton & Co., 2003.

Lichtman, Mitchel. "The FanGraphs UZR Primer". FanGraphs Blog, blogs.fan-graphs.com/the-fangraphs-uzr-primer/. Accessed on November 4, 2020.

Love, Connor. "Cumulative Home Run Ratio and today's home-run hitters". Society for American Baseball Research (SABR), January 2012. sabr.org/latest/cumu-lative-home-run-ratio-and-todays-home-run-hitters/. Accessed on November 22, 2020.

McCracken, Voros. "Pitching and Defense: How Much Control Do Hurlers Have?". Baseball Prospectus, www.baseballprospectus.com/news/article/878/pitching-and-defense-how-much-control-do-hurlers-have/. Accessed on November 8, 2020.

"Park Adjustments." About Baseball Reference, www.baseball-reference.com/ about/parkadjust.shtml. Accessed on November 15, 2020.

Patriot. "Win Shares Walkthrough, parts 1 to 6". Walk Like a Sabermetrician, walksaber.blogspot.com/2005/12/win-shares-walkthrough-pt-1.html to walksaber.blogspot.com/2006/1/win-shares-walkthrough-pt-6.html. Accessed on November 15, 2020.

Patton, Andy. "MLB: Are the Three True Outcomes Killing Baseball?". Unafraid Show, unafraidshow.com/mlb-analytics-three-true-outcomes-killing-baseball/. Accessed on November 8, 2020.

Petriello, Mike. "A new way to measure MLB's best infield defenders". MLB News, www.mlb.com/news/statcast-introduces-outs-above-average-for-infield-defense. Accessed on November 4, 2020.

Petriello, Mike. "Statcast introduces Catch Probability for 2017". MLB News, www. mlb.com/news/statcast-introduces-catch-probability-for-2017-c217802340. Accessed on November 4, 2020.

"Position Player WAR Calculations and Details". About Baseball Reference, www. baseball-reference.com/about/war_explained_position.shtml. Accessed on November 22, 2020.

Reiff, Brian. "A Surprising Win Shares Leader". Bill James Online, www.billjame-sonline.com/a_surprising_win_shares_leader/. Accessed on November 15, 2020.

Reiff, Brian. "Defensive Runs Saved – 2020 Update". FanGraphs Library, library. fangraphs.com/defensive-runs-saved-2020-update/. Accessed on November 4, 2020.

"Retrosheet", Retrosheet, retrosheet.org. Accessed on November 2, 2020.

Ruff, Rivea. "Aaron Judge Seems to Break StatCast with Ridiculous Home Run". Bleacher Report, bleacherreport.com/articles/2723107-aaron-judge-seems-to-break-statcast-with-ridiculous-home-run. Accessed on November 15, 2020.

"Runs created", Wikipedia, en.wikipedia.org/wiki/Runs_created. Accessed on November 2, 2020.

Schott, Thomas E. "Hack Wilson.", Society for American Baseball Research Bio Project, sabr.org/bioproj/person/hack-wilson. Accessed September 20, 2020.

Schwarz, Alan. The Numbers Game: Baseball's Lifelong Fascination with Statistics. St. Martin's Griffin, 2005.

Slowinski, Steve. "Linear Weights". FanGraphs Library, library.fangraphs.com/principles/linear-weights. Accessed on November 2, 2020.

Slowinski, Steve. "OPS and OPS+". FanGraphs Library, library.fangraphs.com/offense/ops. Accessed on September 20, 2020.

Slowinski, Steve. "RZR". FanGraphs Library, library.fangraphs.com/defense/rzr/. Accessed on November 4, 2020.

Slowinski, Steve, "wRC and wRC+". FanGraphs Library, library.fangraphs.com/offense/wrc. Accessed on November 2, 2020.

Slowinski, Steve. "xFIP". FanGraphs Library, library.fangraphs.com/pitching/xfip/. Accessed on November 8, 2020.

Smith, Sean. "What is Zone Rating?". The Hardball Times, tht.fangraphs.com/what-is-zone-rating/. Accessed on November 4, 2020.

Sports Info Solutions, Inc. Fielding Bible. "Overview of the Plus/Minus System". The Fielding Bible Awards, www.fieldingbible.com/overview.asp. Accessed on November 4, 2020.

Sports Info Solutions, Inc. Fielding Bible. "Team Defensive Runs Saved". The Fielding Bible Awards, www.fieldingbible.com/TeamDefensiveRunsSaved. Accessed on November 8, 2020.

"Statcast Outs Above Average Leaderboard". Baseball Savant, baseballsavant.mlb.com/leaderboard/outs_above_average. Accessed on November 4, 2020.

Tango, Tom. "Statcast Lab: Is there a different run value needed based on infield slice?". Tangotiger Blog, tangotiger.com/index.php/site/comments/statcast-lab-is-there-a-different-run-value-needed-based-on-infield-slice. Accessed on November 4, 2020.

Thorn, John and Pete Palmer. The Hidden Game of Baseball (Revised Edition). Doubleday & Company, 1985.

"WAR Comparison Chart". About Baseball Reference, www.baseball-reference.com/about/war_explained_comparison.shtml. Accessed on November 22, 2020.

"WAR Explained". About Baseball Reference, www.baseball-reference.com/about/war_explained.shtml. Accessed on November 15, 2020.

"WAR Explained, Converting Runs to Wins". About Baseball Reference, www.baseball-reference.com/about/war_explained_runs_to_wins.shtml. Accessed on November 15, 2020.

Weinberg, Neil. "RE24". FanGraphs Library, library.fangraphs.com/misc/re24. Accessed on November 2, 2020.

"Win Shares". BR Bullpen. Baseball Reference, www.baseball-reference.com/bullpen/Win_Shares. Accessed on November 15, 2020.

"Win Shares". Wikipedia, en.wikipedia.org/wiki/Win_shares. Accessed on November 15, 2020.

Zimmerman, Jeff, and Dan Basco. "Measuring Defense: Entering the Zones of Fielding Statistics". Baseball Research Journal, Summer 2010, sabr.org/journal/article/measuring-defense-entering-the-zones-of-fielding-statistics/. Accessed on November 4, 2020

The annihilation of records: where have you gone, Babe Ruth?

Chapter outline

The greatest to ever swing a bat

A common comparison of greatness in baseball (perhaps even the ultimate comparison) occurs when a player is linked to George Herman "Babe" Ruth. Ruth is arguably the greatest player to ever swing a bat. His nicknames include "The Bambino, the Sultan of Swat, the Big Bam, the Behemoth of Bust, the Colossus of Clout, and the Maharajah of Mash;" all bring connotations of greatness, of hitting a ball harder or farther than anyone else. Something a mere mortal could not do but also something a player beloved by the baseball gods could do. Boston sportswriter Burt Whitman scrutinized Ruth in July 1918 when he was just becoming an everyday player, writing, "The more I see of Babe and his heroic hitting the more he seems a figure out of mythology."[1] To put this into context, reread this last sentence. Whitman wrote it in 1918 when Ruth was just 23 years old and was beginning his transition away from being an effective starting pitcher.

Sabermetrics. DOI: http://dx.doi.org/10.1016/B978-0-12-822345-1.00006-4

113

There is an old saying regarding sabermetrical studies: *If you conduct a sabermetrical analysis of the greatest players in baseball history and Babe Ruth does not come out at or near the top, then something's probably wrong with your study.*[2] Ruth was the greatest player, without question, for so long, according to so many studies. Is it still true? That old sabermetrical axiom is coming under scrutiny, as players are achieving Ruthian numbers at the plate, threatening to break century-old records.

Ruth's statistics

Open your Internet browser and go to Babe Ruth's page at Baseball-Reference.com.[3] This site gives Ruth's statistics for the 22 seasons in which he played baseball in the major leagues. The first table of data shows Ruth's Standard Batting statistics. Each number that is bold-faced indicates when he led the league in a particular season. Each number that is bold-faced and italicized indicates when he led all major leagues in a particular season. The numbers in italics and gold show career totals and indicate Ruth's all-time career records (in these instances, they are still active records). A quick glance at this table shows how vast Ruth's dominance extended. He was a perennial leader in runs scored (R), home runs hit (HR), runs batted in (RBIs), bases on balls (BB), on-base percentage (OBP), slugging percentage (SLG), on-base-plus-slugging percentage (OPS), total bases, and OPS+, which is the OPS adjusted for ballparks. Ruth's career marks in SLG (0.690), career OPS (1.164), and career OPS+ (206) have led all batters in the majors since Ruth retired in 1935.

According to mlb.com, "OPS+ takes a player's OPS and normalizes the number across the entire league. It accounts for external factors like ballparks. It then adjusts so a score of 100 is league average, and 150 is 50% better than the league average."[4] Ruth is the only player in history to post an OPS+ career value greater than 200. This means he is the **only** baseball player who was more than 100% better than the league average, *over his entire career*! In the history of the game, only 11 major leaguers have attained an OPS+ greater than 206 and each of those were for a single season. Ruth's value is for his 22-season career. In addition, Ruth, as 1 of the 11, bested the 200 OPS+ plateaus 10 times. Furthermore, if we do not include 1914, when he appeared in only 5 games, Ruth posted an OPS+ number always greater than 100, and he had only 3 seasons in which his OPS+ was less than 160. So, what? This means he was always better than the average player, and except for 1916 (when he was primarily a pitcher), 1925 (when he lost much of the season to injury), and 1935 (his final season where he batted 0.181), he was at least 1.6 times better than the average ballplayer. In the year 2021, Ruth will have held the career mark in OPS+ for 100 years.

A hall of fame pitcher?

Besides his prowess at the plate, to which we will return shortly, Ruth was a great pitcher. He broke into the majors as a dominant left-handed starter. Had he spent his playing time only on the mound, he still might have been elected to Baseball's Hall of Fame. In 1921, Tris Speaker, Ruth's 1915 teammate in Boston and fellow Hall of Fame inductee (selected the year after Ruth) is credited with saying, "Ruth made a grave mistake when he gave up pitching. Working once a week, he might have lasted a long time and become a great start."[5] Ruth did give up pitching, and we will never know what could have happened.

Looking again at Ruth's webpage at Baseball-Reference.com, he led the league in different categories while pitching for the Red Sox. As a pitcher, he never had a losing season. He had a 0.671 career winning percentage (94-46, which places him 12th in major league history on the career list[6]) and a lifetime earned run average (ERA) of 2.28. He led all American League (AL) hurlers in 1916 with a 1.75 mark in 40 starts, including 9 shutouts, and completed 107 of 147 starts. Ruth's World Series record was 3-0, with 2 complete games, a shutout, and a 0.87 ERA. He also set the pitching record of 29{2/3}consecutive scoreless innings in the World Series—a record Ruth held for 43 seasons.[7] Furthermore, Ruth's 14-inning victory in Game Two of the 1916 World Series continues to be the longest pitching outing in Major League Baseball post-season history. After Hi Myers' "freak inside-the-park home run"[8] in the first inning, Ruth held the National League Champion Brooklyn Robins (later, Dodgers) scoreless for the next 13 frames; more impressively, he allowed only 5 more hits (4 singles and a double) in those 13 innings. In 1916, the young Red Sox hurler led the AL by allowing only an average of 6.396 hits per nine innings pitched. Furthermore, he did not allow a single home run to any opponent in 323{2/3} innings pitched!

Using a method known as Player Wins, noted statistician Pete Palmer analyzed Ruth's pitching stats.[9] He converted them into runs above average, which was then transformed into wins above average (WAA). He used 10 runs per win. Ignoring the 1914 season (Ruth appeared in a mere four games as a pitcher—he appeared in one other contest as a pinch-hitter), Palmer concluded that Ruth had "the best start of any pitcher through his first 4 years of at least 100 innings pitched per year in the modern era,"[10] which Palmer noted was from 1983 when the pitching distance was moved to its current position. Ruth ranks higher than any other pitcher who had at least 11 WAA in their first 4 years (Palmer found that Ruth had 19.6 wins). There are only 39 such pitchers in history. Yes, better than any other pitcher in his first year, including fellow Hall of Famers Bob Feller, Smokey Joe Wood, Christy Mathewson, Tom Seaver, Bob Gibson, and many other stars!

Ruth once described his pitching with, "As soon as I got out there I felt a strange relationship with the pitcher's mound. It was as if I'd been born out there. Pitching just felt like the most natural thing in the world. Striking out batters was easy."[11]

A home run hitter

The casual baseball fan knows Babe Ruth's name and might consider him as a great home run hitter. He was so much more. However, let's start with the home runs. Ruth hit his first career major league home run on Thursday, May 6, 1915, at the Polo Grounds in New York. The feat took place in a game pitting two modern rivals against each other, the Boston Red Sox and the New York Yankees, before a modest crowd of approximately 5000 spectators. The southpaw Ruth (6-foot, 2-inches tall) was the starting pitcher for Boston, opposed by New York's Jack Warhop, a 5-foot, 9-inch righthander. Ruth was in his first full season in the majors,[12] whereas Warhop was in his last. It was just the fifth game of the season for Ruth, who had a record of 1-0 with a 5.02 earned run average. The *New York Times* affirmed that "the big left-handed pitcher, Babe Ruth, was all that a pitcher is supposed to be, and some more."[13]

In the third inning, "Ruth, who impressed the onlookers as being a hitter of the first rank, swatted a low ball into the upper–tier of the right-field grandstand and trotted about the bases to slow music."[14] Boston had the lead and Ruth was pitching through the New York line-up, prompting the *Boston Daily Globe* to print, "This run looked as tall as the Woolworth Building."[15] The two teams then each scored a few more runs, and at the end of nine innings of regulation the score was tied, so the game proceeded into extra frames. After 12 rounds, Ruth was still on the mound (Warhop had lasted only eight innings). Ruth then "weakened a bit in the 13th, yielding two successive singles, which with a steal gave the much-coveted run to the Yankees."[16] New York had won 4-3,[17] but Ruth's legend as a slugger was born.

For the day, Ruth was 3-for-5 at the plate, including his first home run of the season, raising his batting average to 0.417. Ruth eventually hit 4 HR in 42 games in 1915, and his quartet of over-the-wall hits led the Red Sox.[18] The Babe's 0.315 seasonal batting average was second on the Sox only to teammate Tris Speaker (0.322), but Ruth's SLG (0.576) and his OPS, at 0.952, led the Boston regular players. Remember, he was a pitcher, not expected to be the team leader in batting. Yet if sportswriter Burt Whitman was to be believed (long before Ruth moved to the New York Yankees), Ruth would become a mythical hero in America's national pastime.

How dominant was the Babe with a bat in his hands? As mentioned earlier, in 1918, the 23-year-old Ruth split time between mound duties (winning 13 of 20 decisions, even though he only started 19 of those contests) and being a position player for the Boston Red Sox. He played 59 games in the outfield and 13 more at first base, as Boston Red Sox manager Ed Barrow sought to get Ruth's potent bat into the Red Sox lineup. Never before 1918 had The Babe not been on the mound, but his emergence as a slugger changed the face of baseball forever. In 317 at-bats (AB), he clubbed 11 HR to tie Tillie Walker for tops in the AL. Walker, an outfielder with the Philadelphia Athletics, needed 414 AB to get the same number. A season later, Ruth smacked 29 homers to lead both leagues, setting a new record for HR in a single season, breaking Ned Williamson's 1884 record of 27. Second-most in 1919 was Gavvy Cravath, who hit 12 for the Philadelphia Phillies.

On December 26, 1919, Ruth was sold by the Boston Red Sox to the New York Yankees for an absurd sum of $100,000 (a century later, that figure would be approximately $1.5 million, using the US inflation calculator). On May 1, 1920, Ruth hit his first home run as a member of his new team.[19] It was his 10th game with the New Yorkers and it came off of future Hall of Famer Herb Pennock. In the sixth inning with one out, Ruth, the two-time reigning home-run champion,[20] "lambasted a home run high over the right field grand stand into Manhattan Field."[21] Babe's first home run of the season as a Yankee was labeled a "sockdolager,"[22] or "a heavy, finishing blow,"[23] which was fitting as the ball cleared the roof in right field and virtually sealed the home team's victory. *The New York Times* reported that "the ball flitted out of sight between the third and fourth flagstaffs on the top of the stand. Ruth smashed it over the same place when he broke the world's home run record last season."[24] The only other batter besides Ruth to have hit a ball over the right-field grandstand was Joe Jackson in 1913.[25] It seems somewhat fitting that the first round-tripper for Ruth in a Yankees uniform was such a colossal clout!

In 1920, Ruth's reputation as a home run machine was cemented. Recall what he did in 1918, just 2 years before. He tied the American League lead with 11 HR *for the season*. In 1920, he hit 12 HR in the month of May (he also mashed 12 homers in June and 13 more in July).

In 1920, his first season with the Yankees, Ruth became the first batter in history to hit 30, 40, and then 50 HR in a season, when he clouted 54 round-trippers. In second place that year in the AL was future Hall of Famer George Sisler (29); in the National League, the leader was Cy Williams (15). In 1921, Ruth hit 10 or more HR in 5 straight months (May through September). In 1927, as part of the New York Yankees' famous Murderers' Row, Ruth hit 60 HR to establish a mark that stood for 34 seasons. In September of 1927, The Babe slugged 17 HR in only 27 games! He also drove in 44 runs in that month, on the way to a season total of 165.

Ruth began the season batting fourth in the lineup, in the clean-up spot. On August 6, Manager Miller Huggins moved Ruth to the third spot in the order and that's where he stayed. Who knows? Had Ruth stayed as the Yankees' clean-up hitter, he might have been given number 4, when the Yankees started wearing numbers on the back of their jerseys.[26]

Setting the home run records

What makes Ruth's 1920 totals more impressive is the fact that he hit more HR than every other *team* in the AL (meaning his 54 were more than the totals of each and every other AL team); Ruth also out-homered all but one National League squad. In his career, Ruth out-homered 90 teams, including four ties. Furthermore, he single-handedly out-homered pairs of teams 18 times. Not too many players can claim to have out-homered an entire team, and the last time this was somewhat possible was during World War II when the Chicago White Sox hit a paltry 33 HR in 1943, 23 HR in 1944, and 22 in 1945.[27] Certainly no batter in the past 70 years has out-homered an entire team. When we wish to draw comparisons about dominance, we can compare players with all other players, but often we need to compare them in a relative sense. We will show Ruth's records have stood the test of time, but during his playing days, he was so much better than those around him. Can we say the same for players who have followed? How many Hall of Famers were so much better than those they played with (or against)?

Roger Connor was an infielder for the New York Giants and St. Louis Cardinals who played from 1880 to 1897. Connor hit 138 HR in his career. Ruth had 108 total four-baggers at the end of the 1920 campaign, and he had been an everyday player for only two seasons. Even more impressive, Ruth was sending the ball out of the ballpark; others were "hitting it where they ain't" and racing around the bases; in other words, hitting inside-the-park HR. Not only were Ruth's homers leaving the playing field; many were what would be called tape-measure homers today.

In the bottom of the eighth inning in a July 19, 1921, contest at Detroit's Navin Field, Ruth sent a Bert Cole pitch over the fence at the deepest part of the ballpark.[28] The historic shot officially measured a distance of 560 feet, giving Ruth his 36th home run of the season and the 139th of his career,[29] passing Connor. Ruth, of course, continued to add to his home run total. He hit 23 more round-trippers in 1921, setting a new season high of 59, breaking his own record of 54 set the season before. His career total continued to increase over the next 14 seasons, finally settling on 714 in 1935. That record stood until 1974, when Hank Aaron hit his 715th. Ruth had owned the career home run record for 53 seasons.

On the campus of the University of Tampa (Florida) is a plaque com-memorating what is probably Ruth's longest home run of his career, smashed during a preseason game in 1919. The sign reads:

Babe's Longest Homer

At Tampa's Plant Field on April 4, 1919, "Babe" Ruth, playing for the Boston Red Sox against the N.Y. Giants, smacked a 587-foot home run that set a record in a pre-season game. 4,300 screaming fans saw the feat. Famed Evangelist Billy Sunday, an ex-major leaguer himself, who was conducting a tent revival on the Florida fair grounds nearby, had pitched the first inning of the game, and The Bambino's pace-setting ball was presented to him. Ruth played from 1915 to 1935. He is regarded as the most popular player and greatest slugger in history. One year he hit 60 homers.

One more story about a long-distance Ruth home run. The Yankees traveled to West Point, New York, to play an exhibition game against the Army baseball squad on June 11, 1934, at Doubleday Field.[30] Ruth played two innings both at first base (meaning Lou Gehrig played right field). According to *The Annual Report of the Army Athletic Association, 1929–1935*, "(Cadet William) Priestly laid one in (Ruth's) groove and (Cadet John) Williams retrieved the ball against the tennis court fence."[31] *The New York Times* reported that the blast was "one of the longest and highest ever hit here. It went to deep right center, and although the Babe only jogged lazily to second on it, the crowd understood."[32] The distance was estimated to be 535 feet from home plate. Babe was 39 years-old.

Fellow author Bill Jenkinson credits Ruth with setting home run dis-tance high marks in every ballpark where he played. In an article published in 1996, Jenkinson argues that Ruth "defies rational analysis." Furthermore, "amazingly, many of those records remain unequaled, which is to say that Ruth is a true athletic anachronism. In virtually every other field of endeavor in which physical performance can be measured, there are no Ruthian equivalents. In 1921 alone, which was Ruth's best tape measure season, he hit at least one 500-foot home run in all eight AL cities. There should be no doubt about the authentication of these conclusions. Despite the scarcity of film on Ruth, we can still make definitive evaluations of the approximate landing points of all of his 714 career home runs."[33] We cannot give that same equivalence to other sluggers.

Ruth led the majors in HR 11 times. (In 1930 he hit 49 homers to lead the American League, but Chicago Cubs slugger Hack Wilson smacked 56 to lead both leagues.) In the inaugural All Star Game, in 1933, the 38-year-old Ruth hit the very first home run in the midsummer classic's history. Ruth led the majors 5 times in RBI, 10 times in OBP, 12 times in SLG (and 1 more time he led the AL in SLG) and thus led the majors in OPS 11 times, with two more just leading the AL. Opposing pitchers did not want to face him, so he also led the majors 11 times in getting free passes via the base on balls.

Ruth was more than just a consistent slugger. He batted 0.343 with 271 HR with none on base and 0.352 with 274 homers with men on (not in scoring position); with runners in scoring position, Ruth hit 0.351 with 146 HR. He batted 0.315 in games in which he pitched. He slugged 0.698 at home and 0.682 away. Ruth could have been defined as a clutch hitter. When facing teams with a winning percentage greater than 0.500, Ruth batted 0.352 with an OPS of 1.187 (0.330 against other teams with an OPS of 1.134).

Some split statistics

He had the ability to get better each time he faced a pitcher. Against all starting pitchers in his career, Ruth batted 0.315 in his first plate appearance (PA). This included a 0.456 OBP and a 0.612 SLG (good for an OPS of 1.068). When batting again against that starter, Ruth hit 0.332 in his second PA and 0.346 in his third PA, and, if he saw the starter in a fourth plate appearance (PA), Ruth batted 0.365. Similarly, his OPS increased to 1.091 (second PA) to 1.188 (third PA) to 1.202 (fourth PA). His career OPS continues to lead all other players at 1.1636 for his career. One more thing about OPS: Ruth had six seasons with an OPS of at least 1.250. No other player has ever had even one year with such a mark.

Versus relief pitchers, Ruth had similar success as the game progressed.[34] He batted 0.336 against all relievers in his first PA, 0.354 in his second PA, and 0.350 in the third PA.

In 2019, Allan Wood published an article call "Cool Babe Ruth Facts."[35] In this fascinating addition to Ruth lore, Wood gives data supporting his Ruthian legend often comparing Ruth's feat to others on career lists. For example, Barry Bonds retired in 2007 with a career SLG mark of 0.607; this places him fifth-best, all-time. To surpass Ruth at the top position, Bonds would have had to hit 247 consecutive HR. Or, going back in time, Ruth would have had to have gone hitless in 1147 consecutive AB to drop his SLG below that of Bonds.

Babe Ruth was so consistent, he never went more than two games in his entire 22-year career without a hit or a walk. Many baseball statistics sites, including Baseball-Reference.com, normalize stats over a 162-game season. What did Ruth average? If we look at some stretches of games, we see more of the Ruthian factor.

From April 30, 1923, to August 8, 1924, a span of 250 games, Ruth batted 0.402.[36] Furthermore, from July 24, 1927 to July 30, 1928, a 162-game stretch, Ruth clubbed 71 HR. On August 30, 1923, with a little more than a month left in the season, Ruth was batting 0.405. He finished the season batting 0.393 (205-for-522). If he had collected four more hits over the course of the season (replacing four outs with four hits), he would have joined the 0.400 Batting Average Club (209-for-522).

In 1921, Ruth set the single-season record for most total bases with 457. Two seasons later, he reached base 379 times (via a hit, walk, or hit-by-pitch). This, of course, is still a record.

The Babe finished his career at the top of most offensive categories. In 2503 games, he posted a 0.342 batting average, 10th best in history. His career RBI mark of 2214 stood for 40 years (also broken by Aaron). In 1922 he overtook Dan Brouthers for the highest career SLG (0.696). That number eventually settled to 0.690 (remember, Ruth played until 1935), but it still, almost 100 years later, leads all batters for a career level. Ruth's single-season slugging mark of 0.847 held the top spot from 1920 until 2001 when Barry Bonds posted an 0.863 SLG.

Combining leaderboards

Ruth produced seven seasons with a batting average of at least 0.350, an OBP of least 0.480, and a SLG of at least 0.730. According to Wood, of the more than 20,000 men who have played major league baseball, only five seasons with those numbers exist to a man not named Babe Ruth. Here's the list:

Let's raise the bar. Ruth had three seasons with at least a 0.375 AVG, 0.500 OBP, and 0.750 SLG (from Table 6.1, you can see they occurred in 1920, 1921, and 1923). No one else is in that atmosphere.

Let's return to Ruth's home run reputation. In the five seasons from 1919 to 1923, Ruth hit a total of 218 HR. In that same time frame, only 11 players hit more than 50 HR. Cy Williams was second on the list, clouting half of Ruth's total (109) in the five-season span. Ken Williams hit 108, and no other batter broke the century plateau. Ruth is the only player to ever hit three HR in a postseason game twice. When The Bambino was released by the New York Yankees on February 26, 1935, he had personally hit 28.3% of all of the HR in Yankees history.[37]

Ruth became the first batter to hit 500 HR in a career. From 1930 until 1979, only 12 players had slugged at least 500 career HR (in order of occurrence): Babe Ruth, Mel Ott, Jimmie Foxx, Ted Williams, Willie Mays, Mickey Mantle, Eddie Mathews, Hank Aaron, Ernie Banks, Harmon Killebrew, Frank Robinson, and Willie McCovey. In the next 20 years (through 1999), only four more players joined the list (Reggie Jackson, Mike Schmidt, Eddie Murray, and Mark McGwire). In the next 20 years

Table 6.1 Comparison of AVG, OBP, SLG

Player, Season	Aerage	On-Base Percentage	Slugging Percentage
Babe Ruth, 1920	0.376	0.532	0.847
Babe Ruth, 1921	0.378	0.512	0.846
Babe Ruth, 1923	0.393	0.545	0.764
Babe Ruth, 1924	0.378	0.513	0.739
Babe Ruth, 1926	0.372	0.516	0.737
Babe Ruth, 1927	0.356	0.486	0.772
Babe Ruth, 1930	0.359	0.493	0.732
Rogers Hornsby, 1925	0.403	0.489	0.756
Ted Williams, 1941	0.406	0.553	0.735
Ted Williams, 1957	0.388	0.526	0.731
Barry Bonds, 2002	0.370	0.582	0.799
Barry Bonds, 2004	0.362	0.609	0.812

AVG, average; OBP, on-base percentage; SLG, slugging percentage.

(through 2019), though, there were 11 more newcomers (Barry Bonds, Sammy Sosa, Rafael Palmeiro, Ken Griffey Jr, Alex Rodriguez, Frank Thomas, Jim Thome, Manny Ramirez, Gary Sheffield, Albert Pujols, David Ortiz), giving the 500 Home Run Club 27 members. Unfortunately, many players in this last group have been suspected of tampering with their ability to hit HR.

The cumulative home run ratio

In 2005, Costa, Huber, and Saccoman published a paper entitled, "Cumulative Home Run Frequency and the Recent Home Run Explosion."[38] The authors defined a measure called the cumulative home run ratio (CHR), which is obtained by dividing cumulative HR by cumulative AB. For example, in 1914 Babe Ruth hit 0 HR in 10 AB. This gives the 19-year-old Ruth a proportion of HR to AB equal to 0.00000. The next year (1915), Ruth hit 4 HR in 92 AB; Ruth's CHR over his 2 years as a pro became 0.03922 (4 HR in 102 AB) at age 20. Because he hit 3 HR in 136 AB in 1916, his CHR at the age of 21 is computed to be 0.02941 and so forth. Over his career, Ruth's CHR ended at 0.08501 (714 homers in 8399 AB). Ruth's data are shown in Table 6.2.

Plots of CHR versus player age were included in the paper for the first 20 members of the 500 Home Run Club. As one would expect, at a certain point, the CHR levels out. From Table 6.2, we can see that Ruth

Table 6.2 Cumulative Home Run Ratio for Ruth

Year	Home Runs Hit	At-Bat	Cumulative Home Run Ratio	Age
1914	0	10	0.00000	19
1915	4	92	0.03922	20
1916	3	136	0.02941	21
1917	2	123	0.02493	22
1918	11	317	0.02950	23
1919	29	432	0.04414	24
1920	54	458	0.06569	25
1921	59	540	0.07685	26
1922	35	406	0.07836	27
1923	41	522	0.07839	28
1924	46	529	0.07966	29
1925	25	359	0.07875	30
1926	47	495	0.08056	31
1927	60	540	0.08389	32
1928	54	536	0.08553	33
1929	46	499	0.08609	34
1930	49	518	0.08676	35
1931	46	534	0.08672	36
1932	41	457	0.08690	37
1933	34	459	0.08616	38
1934	22	365	0.08502	39
1935	6	72	0.08501	40

approached a CHR of approximately 0.077 at the age of 26 and for the rest of his career, especially after the age of 30 his CHR hovered between 0.080 and 0.087. He established a mark that many have thought to be unbeatable. Furthermore, his CHR was pretty consistent as his career finished. So, too, were the CHRs of all of the early members of the 500 Home Run Club. Unfortunately, for both Ruth and for baseball, players who attained their 500th home run after 1996 showed interesting trends. McGwire broke Roger Maris' season home run record (61 established in 1961) in 1998 when he hit 70 at the age of 34. Barry Bonds then broke McGwire's record in 2001 when he hit 73 homers. Between 1998 and 2001, Maris' mark was bested six times—once by Bonds, twice by McGwire, and three times by Sosa. What was going on? McGwire's CHR began to steadily increase after the age of 30, from approximately 0.07 to well over 0.090 (at the age of 37!). Barry Bonds and Sammy Sosa's measures also shot up as they aged. Was this natural? In the paper, Costa, Huber, and Saccoman reported that the home run explosion over the past several years "has generated many questions spanning quite a few areas."[39] How can players hit HR with an increasing frequency so relatively late in their careers? Is the pitching that much worse than in Ruth's time? Or Ted Williams' time? On in Mickey Mantle's time? Are other factors involved?

Bonds eventually established a new career home run mark of 762, placing him atop the all-time list even ahead of Hank Aaron. If his record turns out to be tainted, what does that mean for baseball? On a different level, does that reinforce the greatness of Ruth, who established the CHR level without "other factors"? Ruth ranks second all-time with a mark of 11.76 AB per home run. Basically, that means that he swatted one out of the park once in every 12 or so AB (as in, every three games). Mark McGwire broke that record, setting a career value of 10.61 at-bats per home run, but again, were other factors involved? In a story posted online at espn.com on January 11, 2010, McGwire finally came clean, admitting that "he used steroids when he broke baseball's home run record in 1998, but he also said he didn't need performance-enhancing drugs to hit the long ball."[40] Big Mac claimed he was taking the steroids for health purposes. However, did they affect his ability to hit the long ball?

Master of the extra-base hit

Not every one of Ruth's hits were HR, but of his 2873 base hits, 1356 (47.20%) went for extra bases. That is second-best all-time behind Hank Greenberg's 47.97% (although Greenberg had only 1628 base hits). In the earlier days of the national pastime, extra-base hits were known as "long hits." Box scores would separate the long hits into doubles, triple, and HR, but seasonal statistics also had a heading for long hits. Think about it: getting an extra-base hit puts a player's team in position to score a run. Why? If a player hits a double, triple, or home run, he then puts himself into scoring position (and he does score on a round-tripper). Furthermore, if a runner is on base when an extra-base hit is clouted that runner has a great opportunity to score. Let's consider a few examples. In 1921, Ruth belted out 119 extra-base hits (59 HR, 16 triples, and 44 doubles). This is the most ever. Tied for second that season were Ken Williams (of the St. Louis Browns) and Ruth's teammate Bob Meusel with 24 HR. Five players (Boston Braves Ray Powell, St. Louis Browns George Sisler and Jack Tobin, Washington Senator Howie Shanks, and St. Louis Cardinals' Rogers Hornsby) led the majors with 18 triples, so Ruth was close. Cleveland Indians star Tris Speaker led all players with 52 doubles, but Ruth and Hornsby tied for second. Ruth had come close to leading all three categories in the same season. That has occurred only once in history, when Tip O'Neill, playing for the American Association's St. Louis Browns in 1887, led all batters in HR (14), triples (19), and doubles (52). However, that only amounted to 85 extra-base hits.[41]

In fact, Ruth's 1921 campaign was the first season that any player had hit more than 100 extra-base hits. Since then, the feat has occurred 14 more times. Lou Gehrig is the only player to have reached the 100 extra-base hit point twice in his career (117 in 1927 and 100 in 1930), but Ruth hit 99 extra-base hits in both 1920 and 1923 and 97 in 1927. In a leaderboard of percentages of extra-base hits in a single season, Ruth's 1921 value of 58.33% ranks second all-time (Albert Belle had 103 extra-base hits in 1995, out of a total of 173 hits, to give him a ratio of 59.54%). However, Ruth is the only player to have three of the top 10 seasons with highest percentage of extra-base hits (1921 with 58.33%, 1920 with 57.56%, and 1919 with 53.96%).

Ruth and wins above replacement

Using recently developed statistics, Ruth's other career numbers still show dominance. We have already seen that Ruth's adjusted OPS was 206, showing he was 106% better than the league average over his career. One of these stats is called WAR (Wins Above Replacement). It is an attempt to measure the number of wins for which a player is responsible over and above a replacement level player. In fact, if we were to sum up the WAR totals for all players on a team, we ostensibly should obtain a number very close to the team's number of wins. For context, an average player may have a WAR of two, an All Star would be at four, and an MVP candidate in excess of six. When the statistic was first introduced, the replacement player was one who would result in an offensive winning percentage of 0.320. Now, it is around 0.294. This is a very low base; in fact, it means a team of these players would be batting less than 0.300, although this is higher than the mean batting average for each league. Go to the Baseball-Reference.com website for leaderboards for WAR.

Ruth's lifetime offensive WAR is 154.3, a mark that has stood at the top since he surpassed Ty Cobb's 151.2 career number in 1933. The offensive WAR career list names the best of the best. After Ruth and Cobb come Barry Bonds (143.7), Willie Mays (136.8), and Hank Aaron (132.4). Rounding out the top 10 are Ted Williams (126.4), Stan Musial (124.8), Tris Speaker (124.2), Honus Wagner (123.3), and Rogers Hornsby (121.8). Regarding his contemporaries, Ruth led the majors in offensive WAR in 7 seasons (he led the AL in WAR 10 times), finished second in the majors 5 times, and made the top 10 list in every season from 1918 to 1933. How many other players can claim that consistency?

Not surprisingly, Ruth owns the top three "best season" WAR marks in history: 14.1 in 1923, 12.8 in 1921, and 12.5 in 1927. His 1920 mark

(11.8) and 1924 value (11.7) crack the top 10 as well. Looking more closely, we see that from 1920 through 1927, Ruth's peak years, he put up numbers not attained by others. Ruth's overall WAR value (including defense and pitching) is 182.5, again first all-time. He led the AL 10 times and paced in the top 10 17 times. His total bases topped the charts in 6 different seasons.

The Bambino's OBP of 0.4739 ranks second all-time behind Ted Williams's 0.4817, which obviously contributed to Ruth's top OPS value and that might be tops for a long time to come.[42] Ruth did not have the luxury of being a designated hitter, a rule change adopted by the AL in 1973. He played every day, averaging 140 games per season from 1919 through 1933. Toss in 10 postseasons (all in the World Series) in which he averaged 0.326 and hit 15 HR, and Ruth showed he could perform at the highest level anytime, against any opponent.

Breaking Ruth's records

In 1961, the major league season expanded from 154 games to 162 games. When Roger Maris broke Ruth's single-season record for HR, he played in 161 games (incidentally, in 1927, when Ruth hit 60 HR he played in only151 games). After 154 games, Maris had hit 56 homers, and many (including baseball's commissioner) objected and wanted to insert an asterisk in the record books. They did not want Ruth's legacy to be tarnished. Maris hit home run number 60, tying Ruth, in his 158th game. On October 1, 1961, in his 161st and final game of the season, Maris hit number 61 for the new record. Eventually the asterisk was removed, but when McGwire and Bonds closed in on Maris' record, the conversation about breaking Ruth's mark again resurfaced.

Of course, there is an old adage that "records are made to be broken." Roger Maris broke the single-season record for HR. Then, Mark McGwire broke Maris's record, and Barry Bonds broke McGwire's. Hank Aaron broke Ruth's record for career homers, and then Bonds overtook Aaron. One difference between Ruth and the stars of the early 21st century (e.g., McGwire and Bonds) might be the endurance factor. Ruth was dominant for most of his career not just the final few seasons.

Ruth still owns so many career and single-season batting records. When he came along more than 100 years ago, he annihilated the existing records. A few players have emerged in recent generations who showed great potential. Offensive stars such as Mike Trout, Cody Bellinger, Pete Alonso, and Christian Yelich have started their own strings of leading the league in power categories, but only time will tell if they can overthrow Ruth. All might now seem worthy of induction into the Hall of Fame, but even in Cooperstown not all Hall of Famers are created equal. If you ever get an opportunity to visit the National Baseball Hall of Fame and Museum in Cooperstown, New York, you will pay homage to all the greats who have been immortalized with a plaque. Ruth's monument has only 27 words:

Greatest drawing card in history of baseball. Holder of many home run and other batting records. Gathered 714 HR in addition to fifteen in World Series.

This barely took a few lines on the plaque. It specifically lists only two records. What about all those batting records? In recent inductions, players are honored with over 100 words and 10 lines of accomplishments. The font size keeps getting smaller. Any time the player had success, it seems to make it onto his plaque in Cooperstown. It reminds one of the movie *Mr. Baseball*, where New York Yankees player Jack Elliott (played by actor Tom Selleck) tells his manager, "Last season, I led this club in ninth-inning doubles in the month of August."[43] Ruth's performance dominated in so many areas, one might muse how long the plaque would have to be, were he being inducted this year.

Can science explain the stats?

Beyond the statistics, can we scientifically measure how dominant Ruth was as a home-run hitter? In the October 1921 edition of *Popular Science Monthly*, researchers at Columbia University in New York City hooked up Ruth to apparatus after apparatus and "analyzed his brain, his eye, his ear, his muscles; studied how these worked together, reassembled him, and announced the exact reasons for his supremacy as a batter and a ballplayer."[44] This was after his phenomenal 1921 season. One test required Ruth to put a stylus in three holes on a triangular-shaped board in consecutive order. He did it 132 times in 1 minute. Another test required him to press a telegraph key when a light flashed. He responded more than 10% quicker than the average man. According to the article, "the tests revealed the fact that Ruth is 90% efficient compared with a human average of 60%. {Ruth's} eyes are about 12% faster than those of the average human being. {His} ears function at least 10% faster than those of the ordinary man. {His} nerves are steadier than those of 499 out of 500 persons. In intelligence, as demonstrated by quickness and accuracy of understanding, he is approximately 10 percent above normal."[45] The researchers used results from their tests to explain Ruth's superiority. Then, in a surprise, they revealed that he could be even better than his 59-home-run self in 1921. Ruth evidently held his breath while hitting, and "for that reason, he is not getting the maximum force into his batting."[46] The report concluded that by "dissecting the 'home run king' (the researchers) discovered brain instead of bone, and showed how little mere luck, or even mere hitting strength, has to do with Ruth's phenomenal record."[47]

Surely, we will not see pitchers-turned-batters who can put up Ruth's numbers. The game has changed. Some critics might say that Ruth never played at night, never played against African-Americans, never had to battle jet lag, et cetera. He did not get the opportunity to be a designated hitter and did not have to play offense and defense for nine innings every day. However, Ruth had a reputation for playing hard both on the field and off it. He still had to hit the ball where they "ain't" when he stepped into the batter's box, no matter who the opponent was or what he had done the day before. And his success, far and away above those who played before him, with him, and after him, is why we define baseball dominance as Ruthian. As Will Harridge, President of the AL from 1931 to 1959, once said, "To say 'Babe Ruth' is to say 'Baseball.'"[48]

Chapter problems

1. Babe Ruth was primarily a pitcher for the Boston Red teams of 1915 through 1917. Using his seasonal batting averages for these three seasons, extrapolate his statistics to give him 400 AB per season. Preserving his hit-to-walk ratio for the particular season, how many HR would he have added to his career total, had he been a position player instead of a pitcher? What assumptions must you make and what conclusions can you draw?
2. A batter has hit at least 50 HR in a single season 46 times. Ruth has done it 4 times. Calculate Ruth's home run ratio for each of those four seasons (1920, 1921, 1927, and 1928). Do you notice any trends?
3. From September 21, 1928, to July 31, 1934, Ruth had back-to-back games with no hits and no walks only *twice*. That is a span of 786 total games (over seven seasons!). Using his statistics found online at Baseball-Reference.com, calculate Ruth's OBP over this period.
4. Offensive winning percentage is defined as the percentage of games a team with nine of this player batting would win. It assumes average pitching and defense. The top five players in history are Ruth (0.858), Ted Williams (0.857), Barry Bonds (0.815), Rogers Hornsby (0.815), and Shoeless Joe Jackson (0.814). Calculate the winning percentage of the teams which these players played for and compare each player's OWP with his team's actual winning percentage. What conclusions can you draw? List any assumptions.
5. Take a look at the Hall of Fame plaques for inductees from 2015 to 2019. Using the current models as a guide, what would you say about Babe Ruth? How long would the plaque need to be?

Solutions to chapter problems

1. *Babe Ruth was primarily a pitcher for the Boston Red teams of 1915 through 1917. Using his seasonal batting averages for these three seasons, extrapolate his statistics to give him 400 AB per season. Preserving his hit-to-walk ratio for the particular season, how many HR would he have added to his career total had he been a position player instead of a pitcher? What assumptions must you make and what conclusions can you draw?*
Solution: The biggest assumption we have to make is that his ratio of hits to walks (getting on base) remains constant, as does his seasonal home run rate. We assume that he is not better (or worse) than he was in subsequent seasons. We use the equivalence coefficient. From Ruth's statistics, we see that the Babe had 92 AB and 9 BB for a total of 101 PA that resulted in a walk or hit. Hence, if x is the number of additional AB, the following proportion preserves the AB-to-PA ratio:

$$\frac{Actual\ AB}{Actual\ PA} = \frac{Actual\ AB + x}{New\ PA}$$

Remember, we are considering only hits plus walks in PA. So, for Ruth, we obtain $\dfrac{92}{101} = \dfrac{92 + x}{400}$.

To solve this equation for x, we merely cross multiply and isolate the unknown quantity to obtain $x = 272$. In 1915, Ruth would get an additional 272 AB. This implies that he would also receive an additional 27 BB because 92 AB + 272 (additional AB) + 9 BB + 27 (additional BB) = 400 PA.

So, if Ruth was just as good as he always was for these extra 272 AB, then his projected HR total would be: $4 + \dfrac{272}{400}(4) = 6.72$ or

7 HR for the season. We note that the term $\dfrac{272}{400}(4)$ is nothing more than a prorating of the 4 HR Ruth actually hit. We repeat this exercise

for the 1916 and 1917 seasons. For 1916 (when he had 136 AB and 10 BB), Ruth gets additional 237 AB, resulting in additional 2 HR, and for 1917 (when he had 123 AB and 12 BB), he gets additional 241 AB, resulting in 1 additional home run. So, for the 3 seasons, if Ruth's AB increased to 400 per season, he would have hit 6 additional HR. That would give him 720 for his career.

That does not seem like a lot. If we include the $272 + 237 + 241 = 750$ additional AB with his 8399 career AB and multiply them by his career home run total of 714, we find that he hits $714 + \dfrac{750}{(8399 + 750)}(714) = 773$ career HR, meaning that he would have hit 59 HR in those three seasons.

This number would still be the record.

2. *A batter has hit at least 50 HR in a single season 46 times. Ruth has done it 4 times. Calculate Ruth's home run ratio for each of those four seasons (1920, 1921, 1927, and 1928). Do you notice any trends?*

Solution:

Year	At-Bat	Home Runs Hit	Home Run Ratio
1920	458	54	0.118
1921	540	59	0.109
1927	540	60	0.111
1928	536	54	0.101

These rates all indicate that Ruth hit a home run, on average, in every 9 to 10 AB.

3. *From September 21, 1928, to July 31, 1934, Ruth had back-to-back games with no hits and no walks only twice. That is a span of 786 total games (over seven seasons!). Using his statistics found online at Baseball-Reference.com, calculate Ruth's OBP over this period.*

Solution: We include Ruth's hits, walks and hit-by-pitch (reached-on-error is ignored). For the 786 games, Ruth had 2747 AB, 934 hits, and 696 walks. That creates an OBP of 0.593, which is 0.109 more than his career mark.

4. *Offensive Winning Percentage (OWP) is defined as the percentage of games a team with nine of this player batting would win. It assumes average pitching and defense. The top three players in history are Ruth (0.858), Ted Williams (0.857), and Barry Bonds (0.815), tied with Rogers Hornsby. Calculate the winning percentage of the teams which Ruth, Williams, and Bonds played for and compare each player's OWP with his team's actual winning percentage. What conclusions can you draw? List any assumptions.*

Solution: Assumptions: we exclude partial seasons. See Table.

For Ruth, we include 1915 to 1934. From 1915 to 1919, the Red Sox were 423-297-12, and from 1920 to 1934, the Yankees were 1405-895-11.

For Williams, we include 1939 to 1942, 1946 to 1951, and 1954 to 1960 (omitting his war years). The Red Sox were 1446-1168-12.

For Bonds, we include 1986 to 2004 and 2006 to 2007. From 1986 to 1992, the Pirates were 592-540-2, and from 1993 to 2007 (not counting 2005), the Giants were 1178-1023-1.

Player	Offensive Win Percentage	Team's Win Percentage	Ratio
Ruth	0.858	0.605	1.418
Williams	0.857	0.552	1.553
Bonds	0.815	0.531	1.534

Using a ratio of the player's offensive win percentage to his team's win percentage, we find that Ted Williams has the highest ratio, implying that he carried his teams' offense more so than the other two players. Ruth had many more impact teammates than Williams or Bonds. However, who would not want a team of nine Babe Ruths, especially if one of them could pitch?

5. *Take a look at the Hall of Fame plaques for inductees from 2015 to 2019. Using the current models as a guide, what would you say about Babe Ruth? How long would the plaque need to be?*

Solution: Very long (answers will vary). Which statistics are included? Which are left out?

Endnotes

1   Found in Allan Wood, "Cool Babe Ruth Facts," an article in *The Babe*, Bill Nowlin and Glen Sparks, editors, Society for American Baseball Research, Inc.: Phoenix, Arizona, 2019.

2   Attributed to our editor, Gabriel Costa, date unknown (although it's been around since at least the early 1990s).

3   See www.baseball-reference.com/players/r/ruthba01.shtml.

4   See m.mlb.com/glossary/advanced-stats/on-base-plus-slugging-plus.

5   Found online at baberuth.com/quotes. Accessed June 2020.

6   As of the end of the 2019 season.

7   Ruth remains second only to Whitey Ford's 33 2/3 scoreless-innings streak.

8   For more on the game, see Cecilia Tan and Bill Nowlin, "October 9, 1916: Red Sox win Game 2 on a loaned diamond; Babe Ruth goes the distance in 14," sabr.org/gamesproj/game/october-9–1916-red-sox-win-game-2-loaned-diamond.

9   Pete Palmer, "The Babe as a Pitcher," an article in *The Babe*, Bill Nowlin and Glen Sparks, editors, Society for American Baseball Research, Inc.. Phoenix, Arizona, 2019.

10   Palmer.

11   Found online at baberuth.com/quotes.

12   Ruth appeared in five games for the Red Sox in 1914, (two in July and three in October).

13 "New York Clubs Defeat Boston's Two Teams: High and Cook Spill Red Sox in 13th," *New York Times*, May 7, 1915: 11.

14 "Red Sox Lose to Yanks in 13th," *Boston Daily Globe*, May 7, 1915: 9.

15 *Boston Daily Globe*.

16 *Boston Daily Globe*.

17 For more on the game, see Mike Huber, "May 6, 1915: Red Sox pitcher Babe Ruth hits first major-league homer," sabr.org/gamesproj/game/may-6–1915-red-sox-pitcher-babe-ruth-hits-first-major-league-homer.

18 Ruth hit four home runs in 1915 on a Boston squad that only produced a grand total of 14 homers.

19 For more on the game, see Mike Huber, "May 1, 1920: Babe Ruth's first Yankee home run is a 'colossal clout' against Red Sox," sabr.org/gamesproj/game/may-1–1920-babe-ruth-s-first-yankee-home-run-colossal-clout-against-red-sox.

20 Ruth and Tillie Walker of the Athletics tied to lead the American League and the majors with 11 home runs in 1918; Ruth upped his output to 29 to lead baseball in 1919.

21 "Ruth Drives Ball Over Grand Stand," *New York Times*, May 2, 1920: 20.

22 *New York Times*, May 2, 1920.

23 dictionary.com/browse/sockdolager. Accessed August 2017.

24 *New York Times*, May 2, 1920.

25 David Fleitz, "Joe Jackson," SABR Baseball Biography Project, sabr.org. Accessed August 2017.

26 The Cleveland Indians were the first team to wear numbers on the jersey backs. They debuted on April 16, 1929. The Yankees had also planned to wear numbers, but their Opening Day game had been rained out, so it wasn't until April 18, 1929, that fans could see Ruth wearing Number 3.

27 Rudy York led the AL with 34 home runs in 1943. Vern Stephens led the AL with 24 home runs in 1945. Bill Nicholson led NL with 33 in 1944 and Mel Ott had 26. Tommy Holmes led the NL with 28 in 1945 and Chuck Workman hit 25.

28 For more on the game, see Mike Huber, "July 18, 1921: Babe Ruth's 560-ft blast against Tigers sets career home run record," sabr.org/gamesproj/game/july-18–1921-babe-ruth-s-560-foot-blast-against-tigers-sets-career-home-run-record.

29 See baseball-reference.com:8080/players/event_hr.cgi?id=ruthba01&t=b for a log of Ruth's 714 career home runs.

30 For more on the game (and others where Ruth played against the West Pointers), see Mike Huber, *West Point's Field of Dreams: Major League Baseball at Doubleday Field*, Vermont Heritage Press, Quechee, Vermont, 2004.

31 Report of the baseball season of 1934, as published in *The Annual Report of the Army Athletic Association, 1929–1935*.

32 "Yankees Defeat Army Nine, 7 to 0," *New York Times*, June 12, 1934: 30.

33 Found online at bleacherreport.com/articles/2698852-the-longest-home-runs-in-mlb-history. Accessed June 2020.

34 Wood.

35 Wood.

36 Wood.

37 Wood.

38 Gabriel Costa, Michael Huber, and John T. Saccoman, "Cumulative Home Run Frequency and the Recent Home Run Explosion," *Baseball Research Journal*, Number 34, 2005, pages 37–41.

39 Costa, Huber and Saccoman.

40 "McGwire apologizes to La Russa, Selig," an article posted online at www.espn.com/mlb/news/story?id=4816607, January 11, 2010. Accessed June 2020.

41 O'Neill put together of od the greatest seasons, relative to his peers. He also led the American Association in hits (225), runs scored (167), runs batted in (123), batting average (0.435), on-base percentage (0.490), slugging percentage (0.691), OPS (1.180), and OPS+ (213).

42 As of the beginning of the 2019 season, Mike Trout's career OPS mark is 0.9913. The next active player is Joey Votto, 18th on the career list, with a 0.9536 career OPS.

43 See www.youtube.com/watch?v=DgCY_o1SR3o. Accessed June 2020.

44 Hugh Fullerton, "Why Babe Ruth Is Greatest Home-Run Hitter," *Popular Science Monthly*, October 1921, Vol. 99, No. 4: 19.

45 Fullerton.

46 Fullerton.

47 Fullerton.

48 Found online at baberuth.com/quotes. Accessed June 2020.

Steroids, etc.

Chapter Outline

Steroids were excessively available and used by players in Major League Baseball (MLB) basically from the late 1980s to early 2000s, hence we will refer to this time as the "steroids era." The latter part of this period was an exciting time for baseball and fans especially after a strike-shortened season that threatened to end it early. During this time, several players used performance-enhancing drugs of the form of anabolic steroids and/or human growth hormones (HGH), which are all basically drugs that induce unnatural physical gains. For ease of discussion, we will simply call all such drugs steroids, which are illegal in MLB. Strength and size gains from steroid use appear to translate for a batter to hitting more home runs and for a pitcher to more miles per hour (MPH) on their fastball. Although there is little scientific evidence to support either assertion, we do recognize that during this era there were more home runs and more pitchers throwing 90 MPH fastballs than previous years, but whether players did steroids or not, working out by lifting weights was becoming popular, so who is to say this alone wasn't the reason. Reflecting upon my playing days, my joy in doing sabermetrics, and my role as a father to three aspiring athletes, I cannot help but wonder what MLB (and my life) would be like if steroids didn't find the MLB in the early 1980s. This chapter is a general discussion of the effects and impacts of the steroids era on the game today from my perspective.

Home runs are special from my perspective because in one swing of the bat the score of the game can change by up to four runs. Hitting one is the ultimate feat for a baseball player to accomplish in an at-bat and even more so to accumulate over a career. As a kid, I grew up idolizing Jose Canseco, Mark McGwire, Barry Bonds, and the likes of such MLB greats. These players hit a home run nearly every 10 to 15 at-bats, which was amazing to watch especially on ESPN's SportsCenter and its related spin-offs like "Baseball Tonight". These shows provided 24-hour highlights and were an accelerant to profitability and a large part of inducing major change to

Sabermetrics. DOI: http://dx.doi.org/10.1016/B978-0-12-822345-1.00007-6

the game. I like many, watched post-game highlights (home runs) and then went out and tried to replicate what I saw in backyard wiffleball games with my younger brothers. In 1988, I was 12 years-old and a few games into my Little League season when I realized my dream of hitting a home run by sending a long shot over the right-centerfield fence at our park. I vividly recall speedily rounding the bases and passing through the congratulations of my team and jumping into my father's arms. I was nearly in tears and it was only the third inning of the game. As an avid ESPN watcher at that age, I had my first real SportsCenter moment "Dah-Nuh-Nuh." I was fortunate to have a few more of these moments throughout my career.

I took supplements for a little while as a player in college because I thought I needed them (other players were doing it) to get better. I had no idea what I was doing because of the novelty of such efforts during this time. There was no education or program to follow, so like those around me, I just took some white powder (creatine) with water after workouts. We thought it helped achieve greater strength gains because like steroids, creatine when combined with a lifting regime facilitated muscle growth. Although I did get stronger, I didn't love how creatine made me feel (bloated) so I didn't take it often. Statistically speaking, I had one great season and one average season following winter-lifting sessions in which I took creatine. I could not tell you if creatine made me a better player or not, but I am sure that if I signed my late-round contract-offer from the White Sox, I would have been looking for a similar angle to be competitive as I was a sure bubble player. I share this because it's my perception that players during this time, like me, took supplements to gain an edge and often did not consider and/or understand the side-effects.

The steroids era was a great time for baseball and for me. It accelerated the evolution of the game. If we think about baseball at its inception, where players left their gloves on the field, pitchers threw complete games, and hitters simply tried to put the ball in play while broadcasters glamorized play through thoughtful commentary on the radio to kids and families listening outside of the dimensions of the diamond, everything is different now. The game today is more strategic, where managers set line-ups and organize rosters seeking to gain percentage point advantages on possible statistical matchups of power versus power; media moguls induce entertainment flare through ball tracking, highlight reel coverage, and advertising to entice fans to watch. This is fun to watch, This is fun to watch, but unfortunately, we see this impacting youth games today. Teams indoctrinate kids into MLB type performance enhancing regimes as early as the first grade featuring specialized training to maximize power whether pitching or hitting.

Many of the greatest players of this era in MLB did steroids. It's probably also safe to say that most players also did or experimented with steroids. If you believe Jose Canseco's estimates, which he discusses in his myriad of books, nearly 8 out of 10 players were "juicing." Ken Caminiti who sadly lost his life from the side-effects of steroid use indicated a very high prevalence of it as well, but closer to 50%. These two players are MLB's only honest brokers on steroid use. Although both had a few amazing statistical years, they both they had injuries potentially related to their persistent steroid use that limited their longevity.

MLB players used supplements of a variety of forms dating back as far as the 1940s with "greenies," which were uber-caffeine enriched pills used to combat fatigue and gain a mental edge amidst a grueling 165-game schedule. The popularity and usage of these drugs like speed lasted through the early 2000s co-existing with steroids. It's rumored that clubhouses facilitated these addictions whether through souped up coffee elixirs or just general availability like a pharmacy. Taking greenies is just as illegal as steroids and implicitly serves as evidence of MLB's long-lasting culture of supporting illegal supplements in their clubhouses.

Yes, taking steroids is illegal which translates to cheating, but MLB didn't test for it during this time period. Recognizing that the players' union and owners could not agree on testing protocols is short for "We know what's going on and we are good with it." This implies that players had equal access to the drug and used it at their discretion. Regardless, it's hard to understand that given a pervasive problem with concerning side-effects for players, MLB did not enact at the very least rehabilitative and counseling services for players that used steroids to help them manage the side-effects of it. This ultimately is the integrity-flaw in MLB. It's not so much that players were trying things to gain a competitive edge, because this happens in every sport, but that owners and managers enabled it despite the severe health risks associated with steroid usage. The commissioner and owners protect the integrity of the game and the players. In this case they failed to do both.

When it became apparent to the media that players were taking steroids, MLB ignored it. Rather than take ownership of the organizational culture that allowed and even encouraged steroid use, MLB outsourced the investigation to former Senator George Mitchell in 2006. Mitchell's report indicated the pervasiveness of steroids but failed to blame MLB ownership. Although identifying roughly a hundred players as testing positive for steroids or using them by trainers or team affiliates over a multi-year period, cooperation by players was basically zero, which aligned with the players' union stance on the matter. Senator Mitchell however suggested not punishing these players formally, which Selig didn't, but the commissioner's lack of ownership and his role in the results of the report allowed the media to write a steroids' narrative of their own that persists today and is keeping the great players of this generation out of the Hall of Fame, labeling them as cheaters.

Over the course of baseball history, there was a charming narrative of player uniqueness that endeared them to fans. Some players had quirky personalities, superstitions, tendencies, while others had weird swings, crazy pitching motions, curly mustaches, or funky running styles. Managers made "gut-calls" and emotional decisions like sending Kirk Gibson in to pinch hit for the Dodgers in the World Series when he could barely walk! This is what stories were made of and what made a day at the ballpark magical for fans. Vin Scully laments that baseball was not made for TV, but the radio was its calling for these reasons. MLB is a for-profit organization and over the course of the steroids era, positioned itself and individual franchises to be very profitable through lucrative media deals. The timing of the home run barrage of the steroids era especially McGwire versus Sosa was perfect for ESPN's and other media moguls' business models that were just starting to feast off the power surge in MLB. As a result, player contract prices soared. Small market teams turned to analytics and things like radar guns, statistics, and videos to optimize their rosters and build teams to compete with the likes of the Yankees and Red Sox. Eventually, small market analysis transitioned through the league and led all teams to implement some form of analytic analysis in their operations.

Considering all these changes or rather improvements to players, equipment, and analysis, MLB still maintains the same rules, field dimensions, and balls as it has for years. After Tiger Woods launched his assault on the PGA tour, courses began Tiger-proofing their course, because he made it look to easy to play them and the PGA Tour feared no other player could win. Thus, golf courses changed and made longer holes with tougher greens while placing restrictions on club and ball technology. To his credit, Tiger was probably the first golfer to advertise his commitment to performance enhancing activities like weightlifting leading many players to follow his lead. This focus on building strength across the tour led to a general power assault on golf with most players averaging over 300-yard drives, which was a feat unheard of when Sam Snead was winning tournaments. The popularity and success of physical training on performance naturally led players toward personal goal-setting and research that set exceptional feats, like throwing 100mph or hitting a 450' home run, as interim goals. MLB started inculcating physically enhancing activities into practice regimes that are now also being included in statistical analysis and performance optimization efforts.

Baseball (sabermetricians) admires the anomalies. The "freaks" who dominate the game relative to the competition. We often translate domination as hitting the most home runs or throwing 100 MPH fastballs. Heck, we are still talking about Babe Ruth like he played last year. Many consider him the greatest player ever. Statistically speaking Ruth is the greatest outlier ever (relative to his competition). While statistics during his era pale in comparison to today's, Ruth still ranks in the top 10 of most offensive categories today; however, if you think Ruth could play in today's game you are just wrong. Watching videos of his training regime with Art McGovern demonstrate that his actions are admirable for his generation, but laughable by today's standards. Ruth does credit his late career turn-around to such physical training, which he did publicize. Surprisingly, more players did not follow because many still believed that lifting weights and gaining muscles made you tight and that this tightness prevented you from having a loose and fluid swing or throwing motion.

There are many articles and books on player performance and involvement with steroids. Some attempt to quantify the impact of steroid use on a player's statistics. Most of the research focuses on batter performance although we know pitchers were prevalent steroid users as well. The difficulty of this type of analysis to make conjectures is that there has never been an actual experiment done to identify or isolate the effects of steroids on a player's baseball skills never mind one that looks solely at in-game performance. Given the illegality of steroids, there are no players that openly tell the truth about their use (except Canseco and Caminiti), which makes analyzing or isolating its effects impossible. Players take steroids in cycles often combined with intensive workout regimes. We do not know how many cycles a player needs to do to be successful. Regardless, there are some exceptional efforts to make inferences, but none of them have found anything noteworthy. I can summarize the main findings of this literature in two statements: because (1) more players hit more than 40 home runs, and (2) more players hit more home runs later in their career than previously during this era than previous years on order of magnitude of 10 or 20 players, steroids is the reason. A few authors indicated pitchers could throw 1 or 2 MPHs harder from taking steroids, but then again, the applicability of the data is questionable given varying collection efforts and the general idea that all players were doing something performance enhancing like weight training.

Fig. 7.1 depicts the general trends, reflective of a variety of sources of data, of the steroids era graphically in which we see things like bunts, steals, "balls put in play," and complete games decline, while power numbers like home runs, size, and injuries rise.

Current sabermetrics efforts that apply advanced statistical techniques should stop looking to compare players of today with the players of old and instead focus on what the future of the game may look like considering these trends. Currently, teams are not willing to sacrifice outs for the opportunity to hit a home run. I see a potential for MLB rosters to expand as pitcher and player specialization will likely continue and lead most to find a niche in the game, like pitching the 3rd inning or playing a certain position of a shift (Fig. 7.2). We will continue to see strength on strength in nearly every scenario like World War I trench warfare unless something changes.

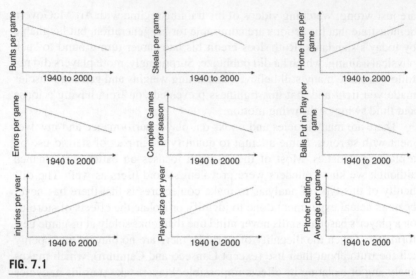

FIG. 7.1

Graphical Depiction of general trends of the 1940s Through the Steroids Era.

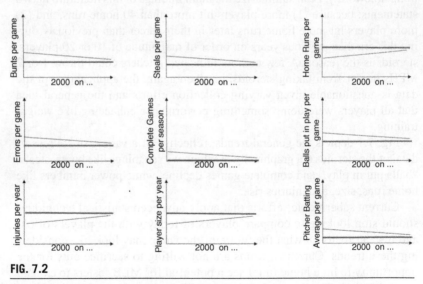

FIG. 7.2

The future of Major League Baseball? Anticipated plateau values in red of key metrics given the current rate of physical development and MLB strategies.

A result of this power surge is that kids don't play stick ball anymore because being creative and having fun isn't going to lead them to the pro's. I say this because as early as 7 years old kids begin youth programs that include spring training and MLB type tournament schedules with 40 to 60 games a summer. When my son was in the second grade, we turned down an "offer" from a team with such regimes. We said 'no,' which I didn't realize at the time meant he wouldn't play baseball anymore. The problem with saying no teams like this whether you agree with it or not, it's that there are no other options to play. When all the top youth players leave for club teams, there is no one left to play locally. Although the media makes baseball accessible to many, the trickle-down effects of these youth programs selecting players at early ages makes the game less accessible. There doesn't seem to be an ambassador of the game seeking to maximize the number of kids playing baseball or at least one that isn't linked to profit margins. I realized this after the baseball episode and now I am in the middle of this youth sports vortex with all three of my kids in all sports. We spend thousands of dollars playing sports, doing private lessons, and traveling to other states and even countries to play. It's a lot of pressure on both parent and kid and I am constantly worried about being priced out of opportunities for my kids.

Research indicates that excessively repeating the same motion leads to injuries. It's hard to analyze injuries in the MLB throughout history because of how they report players using the injured reserve or injury list but given the number of trainers on a team and efforts to count and track player reps it's obviously a concern especially given lucrative player salaries. The problem isn't so much when players get to the MLB because the resources exist there for rehab and treatment rather it's in the prior years. Yes, children who specialize in just one sport are more susceptible to injury than if they were playing multiple sports, which appears to be a dated concept. The MLB season in and of itself is crushing for a player upon arrival, but now players start such activities at age 7 so they arrive to the league with 10 to 13 years of this type of wear and tear on their bodies. It's only getting worse as kids seek the services of speed coaches, powerlifting coaches, and/or even mental to gain an edge on the competition. The resultant is that players and teams are basically equal forcing coaches to experiment using players in creative ways to gain a tactical edge. In lacrosse, ball possession is so important because now teams are so similar in skill that the team with the most possessions wins. There is a position called the Face-Off and Get Off aka the FOGO who just does face-offs, i.e., they do the face-off and leave the field whether they get the ball or not. These players don't even get to play the game. Additionally, the training for such positions is incredibly intensive and position-specific, in which nothing transfers over to the field game, making these players completely one-dimensional and reliant on current rules.

Conclusion

Baseball needs to get with the times, that is, evolve too. The game lags the evolution of the players, equipment, and media and analytical technologies. It's just now starting to use player shifts and experiment with relief pitching plans so it's only a matter of time until we see player specialization like we do in lacrosse. Even the NFL has come around to make some changes by increasing the distance of extra-point kicks after touchdowns.

There is no statistical comparison between players starting in the steroids era to players of prior years because they got too big and strong too fast to document it. The technology wasn't available to quantify the change as it was occurring like it is now. Today's players would crush the players of old. Have you been to the hall of fame and seen the uniform differences? Players today could not even get those jersey's over their heads. Come on baseball wake up, your players are too big and strong for your sandboxes and playpens! You are creating a game that limits accessibility and fun, no one wants to watch WWI on the diamond albeit even from a luxury box.

Yes, these trends appear inevitable given physical condition of players and technology improvements in bat, glove, ball design, but clearly prevalent steroid use accelerated progress.

Scandal scarred: a discussion of our national pastime's controversial history

8

"Hall of Famer Selig, in money-first silent confederation with the MLBPA, allowed players to become so infested with record-busting, drug-swollen sluggers they were dragged before Congress to further, under oath, skirt the truth through the inability to remember, or in the case of Sammy Sosa forget he could speak English"[1]

Chapter outline

Bud Selig, Sammy Sosa, and the rest of the steroid-stained era did monumental damage to the game we all love. The statistical impact alone is enough to make a case that the period defines the greatest scandal our national pastime has ever endured. But was it? Scandal and controversy have joined forces to permit us spirited discussion and endless debate about whether Rose should be in Cooperstown, why very few know of Cool Papa Bell, or whether the 1972 trade between two Yankees that swapped wives (and kids and dogs) was, as it has been labeled, "The Trade of the Century."[2] What is the greatest scandal? Which controversy ranks higher than all the

Sabermetrics. DOI: http://dx.doi.org/10.1016/B978-0-12-822345-1.00008-8
143

rest? We will go through nine innings of madness, providing some background, specifics, and an occasional opinion. Then we will leave it to you to decide where the Astros' sign stealing caper and the Black Sox chicanery rank in the pantheon of big-league baseball scandal.

First inning: does the Hit King's Hall of Fame punishment fit the crime?

Pete Rose was one of my boyhood heroes. A key cog in the Big Red Machine, Charlie Hustle could do it all (even bunt - Rose's career bases-empty-bunt hit percentage is top 10 all-time in major league baseball history with 65 base hits in 93 attempts[3]). He played hard (ask Ray Fosse), never backed down, and was a central figure on the best baseball team of my youth. "The Big Red Machine was the most entertaining team ever," Rose said in 2015. "We had Gold Glove guys up the middle with Davey (Concepcion), Cesar (Geronimo), Johnny (Bench), and (Joe) Morgan. We had batting title type players like me. We had RBI guys like (Tony) Perez… we had personality."[4] With Sparky Anderson at the helm, Rose represented the best of that collection of Hall of Famers and he was always fun to watch (except when he wasn't, like in Game 3 of the 1973 National League Championship Series… I'm a Met fan… or when he was insisting that he never bet on baseball when it was becoming increasingly clear that he did). So that brings us to our first point of discussion. We all know why Rose is not in the Hall of Fame, but should he be enshrined? The number that defines Rose's greatness—4,256—does not tell his whole story.

It is important to draw a distinction between the question of whether Rose deserves a place in Cooperstown and whether Rose's ban for life from the game should be lifted. On August 24, 1989 the Associated Press (AP) reported that "Cincinnati Reds manager Pete Rose, one of the greatest players in the history of baseball and holder of 19 major league records, was banned for life today for betting on his own team—a charge he continued to deny". The AP report identified Rose as the 15th person banned for life in baseball history and the first since 1943. Ominously, and perhaps prophetically, the report indicated that "No one banned for life ever has been reinstated". The AP report was clear that there was no mention in the agreement between Rose and Commissioner Bart Giamatti that outlined the specifics of the ban or whether the suspension would keep Rose from being inducted into the Hall of Fame. It did, however, quote Rose conceding that it was "a very sad day" but clearly over-optimistically asserting that he was only "going to be out of baseball for a very short period of time."[5] That brief period has now reached more than 30 years with no end in sight.

After Giamatti died on September 1, 1989 (only 8 days after banning Rose for life), Rose applied for reinstatement in 1992 and 1997 but

Giamatti's successors, Fay Vincent and Bud Selig, left the ban in place.[6] On January 25, 2015 Rob Manfred assumed duties as the 10th Commissioner of Major League Baseball. Less than 60 days later, Manfred received a formal reinstatement request from Rose and said, "I'm prepared to deal with that request on its merits". Good news for the Hit King? Perhaps, but even more encouraging (at almost 75 years old, Cooperstown was Rose's priority) was Manfred's contention that "Rose could still be elected to the Hall of Fame without being reinstated by MLB because the Hall is not operated by MLB."[7] But it was not that simple. In February 1991, the Hall amended its eligibility rules just months before Rose was supposed to appear on the ballot for the first time. The Hall of Fame "decided then that any player on baseball's ineligible list would be ineligible for the ballot."[8] So while Manfred was technically correct that Charlie Hustle could be elected without being reinstated by Major League Baseball (predicated on the Hall reversing its own amendment to the eligibility rules), the reality was that the two were inextricably linked and Rose wasn't going to be enshrined without a favorable decision from the new Commissioner. Perhaps the real story here is that Manfred sidestepped the question by saying the Hall can do what it wants, and the Hall evaded the issue by tying ballot access to the ineligible list. Rose is caught in the middle—no one to whom he can appeal yet nobody taking responsibility for keeping him out.

Complicating things for Rose was former MLB Special Counsel, John Dowd, who investigated Rose in 1989. Dowd advocated maintaining the lifetime ban when Rose made his formal request to the newly appointed Manfred in the spring of 2015, telling the Cincinnati Enquirer "Pete committed the capital crime of baseball. But this is bigger than Pete Rose. There is a reason we haven't had another gambling case in 26 years. This case wasn't about Pete—this case was about protecting the integrity of the game."[9] Things got worse for the Hit King that summer, prior to Manfred's decision on his reinstatement request, when ESPN's *Outside the Lines* revealed pages from a bookmaker's betting notebook that "seemed to prove what those who have looked closely at Pete Rose already all but knew: that in addition to betting on baseball while managing the Reds, he bet on the game while playing too."[10] Betting records kept by Rose's friend, Mike Bertolini, were available to Dowd back in 1989 but were not widely seen until they were leaked into the public domain in June of 2015. ESPN reported that Bertolini's notes indicated that "Rose was betting thousands of dollars on games at least as early as 1986, his last season as a player-manager."[11]

Six months after the ESPN revelation and almost 30 years after Greg Minton surrendered #4,256 in a 2-0 loss to Rose's Reds in Cincinnati, Manfred seemed to slam the door shut on Rose's reinstatement (and thus, on ever entering Cooperstown). As New York Times columnist Michael Schmidt wrote on the day of Manfred's decision to leave in place the permanent ban on Rose, "The player with more hits than anyone else in the sport's history will continue in an exile that has long kept him out of the Hall of Fame." Why? As Schmidt quoted the Commissioner, "In short, Mr. Rose has not presented credible evidence of a reconfigured life either by an honest acceptance by him of his wrongdoing... or by a rigorous, self-aware and sustained program of avoidance by him of all the circumstances that led to his permanent ineligibility in 1989."[12]

Rose's exile and his exclusion from Hall of Fame consideration continues to be the source of much debate. Many who are already part of that exclusive club believe Rose belongs in Cooperstown. Fellow Reds alumnus and current Hall of Famer Barry Larkin does not equivocate. "He has 4,256 hits. Period."[13] Cal Ripkin, the Iron Man and 2007 Hall of Fame inductee agrees. "Petey, when you look at his career, he's one of the game's best players... Yes, he should be in the Hall of Fame."[14] Others start with "4,256 reasons" but go on to highlight that Rose also owns the records for most career games (3,562), most career at-bats (14,053), and most seasons with 200 or more hits (10, subsequently tied by Ichiro Suzuki in 2010). In addition to those well-known records, Rose's versatility was unparalleled; Charlie Hustle is also the only player in MLB history to have played 500 games at five different positions (1B, 2B, 3B, LF, and RF).[15] If one can decouple Rose's lifetime ban from the Hall of Fame question, the debate ends before it even begins.

Unfortunately, the debate did begin, the moment the Hall of Fame changed the eligibility rules in 1991 just as Rose was about to appear on the ballot for consideration by the Baseball Writers' Association of America. After Manfred denied Rose's request for reinstatement in December of 2015, Rose and his lawyers opted to pursue another route to Cooperstown. In a letter to the President of the National Baseball Hall of Fame dated September 27, 2016, Rose's attorneys took exception to the timing and substance of what has become known as the 'Pete Rose Rule' (which makes those on the MLB ineligible list not eligible for consideration into the Hall). "No one associated with the game other than Pete has ever been categorically denied eligibility from day one after the conclusion of his career for actions having nothing to do with the way they played baseball" wrote attorneys Raymond Genco and Mark Rosenbaum. Genco and Rosenbaum went on to point out that when Rose and Giamatti culminated the settlement agreement in 1989 "it was well understood that Shoeless Joe Jackson had been eligible for the Hall of Fame for the full duration of his eligibility" (and in fact had received some votes for admission to the Hall of Fame over the years).[16] The ostensible movement of the goalposts on the eve of Rose's name going onto the ballot is a compelling point for those who believe he should be in the Hall of Fame. So, I ask you again: Does the Hit King's Hall of Fame punishment fit the crime?

Second inning: did major league baseball get the all-century team right?

One of the best things about baseball is the opportunity that its rich history offers for spirited debate. Our First Inning discussion surrounding Pete Rose, his ban from baseball and the Hall of Fame is complete. However, myriad disagreements about the sport abound. The 2020 pandemic rules notwithstanding, should the National League adopt the designated hitter or does the American League continue to diminish the game by keeping pitchers out of the batter's box? Are 21st century "by the book" pitching changes using designated 7th and 8th inning relievers to set up a team's closer, or even the more recent use of analytics and high leverage opportunities, smart tactics? Or… do those updated approaches remove natural instinct (common sense?) from managerial decisions to the detriment of the game? With home runs up 32 percent and strikeouts up 67 percent over the last 40 years,[17] has the recent emphasis on launch angle and the resulting record of low balls put in play been responsible for dwindling interest in America's pastime? Answers to these and other controversial topics typically stimulate strong emotions in debate, and rightfully so. There are legitimate arguments supporting both sides of these and other important questions. There is one issue, however, that necessitates no debate or discussion. From my perspective, Major League Baseball simply got it wrong when it named the MLB All-Century Team in 1999.

At the turn of the century, an "expert" panel compiled a list of the 100 best players since 1900, and from that list fans voted for the "greatest players" of the 20th century. As a result of the vote, the top two vote getters from each infield position along with the top nine outfielders and top six pitchers were placed on Major League Baseball's All-Century Team.[18] One need not go any further than first base to recognize that MLB's "All-Century Team" was flawed. While the Iron Horse was rightfully identified by MLB as one of the two First Basemen, Jimmie Foxx was left off in favor of Mark McGwire. Seriously? Let us look at the numbers and you decide which player was more deserving of All-Century Team status (all numbers represent career statistics):

BA:	Foxx—.325	McGwire—.263	(Foxx plus .062)
HITS:	Foxx—2646	McGwire—1626	(Foxx plus 1020 hits)
HR:	McGwire—583	Foxx—534	(McGwire plus 49 home runs)
RBI:	Foxx—1922	McGwire—1414	(Foxx plus 508 RBIs)
OBP:	Foxx—.428	McGwire—.394	(Foxx plus .034)
OPS:	Foxx—1.038	McGwire—.982	(Foxx plus .056)
WAR:	Foxx—93.9	McGwire—62.2	(Foxx plus 31.7)

Without even addressing the possible impact of steroids on McGwire's career numbers, Jimmy Foxx was far more productive offensively. Given that defensive metrics were very similar (McGwire committed 103 errors in 1,763 games at first base with a fielding % of .993 while Foxx committed 155 errors in 1,919 games at first base with a fielding % of .990),[19,20] did MLB get it right putting McGwire on the All-Century Team at Foxx's expense? How about the rest of the roster? To be fair, it should be noted that MLB justifiably added Warren Spahn, Christy Mathewson, Lefty Grove, Honus Wagner, and Stan Musial after they were originally left off the All-Century Team. It is difficult to argue with any of those additions, but if MLB decided to allow for subjectivity and additions after the fact, what about Foxx? How about Roberto Clemente, who many fans argue should be there before Ken Griffey Jr.? Here is a look at the numbers of those who made the cut, organized by votes received and inclusive of critical career statistics. Take a look and then answer the question: Did MLB get the All-Century Team right?

NAME	POSN	VOTES	WINS	LOSSES	ERA	IP	Ks	WHIP	WAR
Ryan, N.	Pitcher	992,040	324	292	3.19	5386	5714	1.247	81.3
Koufax, S.	Pitcher	970,434	165	87	2.76	2324	2398	1.106	48.9
Young, C.	Pitcher	867,523	511	315	2.63	7356	2803	1.130	163.8
Clemens, R.	Pitcher	601,244	354	184	3.12	4916	4672	1.173	139.2
Gibson, B.	Pitcher	582,031	251	174	2.91	3884	3117	1.188	89.2
Johnson, W.	Pitcher	479,279	417	279	2.17	5914	3509	1.061	164.5
Spahn, W.	Pitcher	337,215*	363	245	3.09	5243	2583	1.195	100.1
Mathewson, C.	Pitcher	249,747*	373	188	2.13	4788	2507	1.058	106.0
Grove, L.	Pitcher	142,169*	300	141	3.06	3940	2266	1.278	106.7

NAME	POSN	VOTES	BA	HR	RBI	HITS	OBP	OPS	WAR
Bench, J.	Catcher	1,010,403	.267	389	1376	2048	.342	.817	75.2
Berra, Y.	Catcher	704,208	.285	358	1430	2150	.348	.830	59.5
Gehrig, L.	1B	1,207,992	.340	493	1995	2721	.447	1.080	114.1
McGwire, M.	1B	517,181	.263	583	1414	1626	.394	.982	62.2
Robinson, J.	2B	788,116	.311	137	734	1518	.409	.883	61.7
Hornsby, R.	2B	630,761	.358	301	1584	2930	.434	1.010	127.1
Schmidt, M.	3B	855,654	.267	548	1595	2234	.380	.908	106.9
Robinson, B.	3B	761,700	.267	268	1357	2848	.322	.723	78.5
Ripkin Jr., C.	SS	669,033	.276	431	1695	3184	.340	.788	95.9

NAME	POSN	VOTES	BA	HR	RBI	HITS	OBP	OPS	WAR
Banks, E.	SS	598,168	.274	512	1636	2583	.330	.830	67.8
Wagner, H.	SS	526,740*	.328	101	1732	3420	.391	.858	130.8
Ruth, B.	Outfield	1,158,044	.342	714	2214	2873	.474	1.164	182.5
Aaron, H.	Outfield	1,156,782	.305	755	2297	3771	.374	.928	143.1
Williams, T.	Outfield	1,125,583	.344	521	1839	2654	.482	1.116	121.9
Mays, W.	Outfield	1,115,896	.302	660	1903	3283	.384	.941	156.2
DiMaggio, J.	Outfield	1,054,423	.325	361	1537	2214	.398	.977	79.1
Mantle, M.	Outfield	988,168	.298	536	1509	2415	.421	.977	110.2
Cobb, T.	Outfield	777,056	.366	117	1944	4189	.433	.944	151.0
Griffey Jr., K.	Outfield	645,389	.284	630	1836	2781	.370	.907	83.8
Rose, P.	Outfield	629,742	.303	160	1314	4256	.375	.784	79.7
Musial, S.	Outfield	571,279*	.331	475	1951	3630	.417	.976	128.3

*Added to All-Century Team by committee after not garnering enough initial votes.[18,20]
ERA, Earned run average; OBP, on-base percentage; OPS, on-base percentage plus slugging; POSN, position; WAR, wins above replacement; WHIP, walks and hits per inning pitched.

Third inning: was this the trade of the century?

Now that the reader has considered the best ballplayers of the 1900s, it is time to consider what has been dubbed the "Trade of the Century". This qualifies more as scandal than controversy, but regardless of how you interpret it, the transaction arising from the game's most storied franchise elicited strong emotions, discussion, and debate across the country. On March 6, 1973 Phil Pepe of the New York Daily News reported that Yankee pitchers Fritz Peterson and Mike Kekich disclosed that "they had exchanged families... Peterson moving in with Susanne Kekich and her two daughters, Kekich moving in with Marilyn Peterson and her two sons".[21] Wait...what? That's right, "... it was a scandal of epic proportions, the likes of which had never been seen before"[2] but Fritz Peterson and Mike Kekich did indeed exchange everything including wives, homes, kids, and even dogs. Pepe reported that according to Kekich, "We didn't swap wives, we swapped lives."

In the early 1970s, divorce in the United States was still relatively rare and wife swapping (or life swapping) was only really a thing out in California. Kekich and Peterson were obviously aware of the unfavorable impression their "trade" might have on fans, the media, and society in general. Pepe's Daily News report indicated that Kekich insisted "unless people know the full details, it could turn out to be a nasty type thing" and that Peterson "hoped that you won't make anything sordid about this". The pitchers wanted to spin it, sordid as it was, especially given the social norms that were prevalent 50 years ago. Peterson, who remains married to the former Susanne Kekich, held his ground on the appropriateness of the transaction even years later, telling a reporter in 2013 that "It's a love story. It wasn't anything dirty."[22]

Unfortunately for Mike Kekich and the former Marilyn Peterson, the story did not have a happy ending. They have parted ways, and the two ballplayers who were once roommates on Yankee road trips have not spoken in years.[2] Kekich, in an interview long after the swap, offered a hint at his frustration and sense of betrayal. "All four of us had agreed in the beginning that if anyone wasn't happy, the thing would be called off," Kekich said, "but when Marilyn and I decided to call it off, the other couple already had gone off with each other."[23] Like with all trades in baseball, one side usually wins while the other is left to wonder "What if?" So, the reader is asked now, what if the Mets had not traded Nolan Ryan to the Angels for Jim Fregosi on December 10, 1971? Would Peterson for Kekich displace that Metropolitan disaster as the Trade of the Century?

Fourth inning: was segregation the greatest MLB scandal of all?

James Thomas "Cool Papa" Bell was fast. OK, you are thinking, but a lot of ballplayers are fast. Maybe, but apparently not Cool Papa Bell fast. "One time he hit a line drive right past my ear," Satchel Paige once said. "I turned around and saw the ball hit his rear end as he slid into second." Not convinced? Stories of Bell's base running speed abound—advancing two bases on a bunt or beating out comebackers to the mound. Paige once even told a story of rooming with Bell on road trips. "Cool Papa was so fast that he could flip the light switch and be in bed before the room got dark."[24]

Cool Papa Bell and Satchel Paige were only two of hundreds of enormously talented baseball players who were disgracefully denied the opportunity to compete in the Major Leagues solely based on the color of their skin. This scourge of baseball history remains a stain that will never be erased, but thanks to the efforts of courageous men and women, starting with Jackie Robinson and Branch Rickey, Paige ultimately went on to pitch for the Cleveland Indians, St. Louis Browns, and Kansas City Athletics. Alas, no such good fortune visited Cool Papa, who finished his career with the Homestead Grays in 1946, the same year Robinson brought the color barrier down when he made his first appearance with MLB's Montreal Royals in the Triple-A International League.[25] Despite being shamefully denied the opportunity to compete in Major League Baseball, Cool Papa Bell nevertheless still spoke for all of us when he said, "Because of baseball, I smelled the rose of life".[26]

Bell's extraordinary professionalism, class, and dignity are noteworthy but why were he and the rest of our baseball brethren denied the opportunity to compete in the major leagues just because of, as Rickey put it in a banquet speech, the pigmentation of a man's skin? In his address to the *One Hundred Percent Wrong Club* (has there ever been a more apropos name for those who barred blacks from baseball?) in Atlanta, Georgia on January 20, 1956 Rickey called America the "Cradle of Liberty,"[27] but until he and Robinson broke the color barrier MLB was anything but the land of the free.

Kenesaw Mountain Landis is rightly identified as a prime offender of perpetuating the color barrier in baseball, but it started long before Landis assumed duties as MLB's first Commissioner on November 12, 1920. Following the Civil War, The National Association of Base Ball Players (NABBP—the sport was spelled with two words in the 19th century) was formed in 1867 and shamefully banned black players from playing in the association. Though some rosters did contain "a few African-American" players for a brief period of time, by 1890 a "gentleman's agreement" was made, tacitly banning black players from ever playing in the Major Leagues. By the turn of the century, this color barrier was firmly entrenched at the highest level of America's pastime.[25]

Following the 1919 Black Sox scandal (more on that in the Sixth Inning), organized baseball's governing body, the National Commission, decided to disband and reorganize, naming MLB's first ever Commissioner. Recruiting federal judge Kenesaw Mountain Landis, the commission called on Landis "to come in and clean house as Commissioner. They gave him absolute power…".[28] While Landis successfully restored credibility and public confidence in the game with his strong response and punishment of the eight offending members of the Chicago White Sox, the argument simply cannot be made that Landis restored the integrity of the game. Yes, the Commissioner decisively dealt with the gambling issue and adequately resolved that 1919 scandal. However, by preserving the insidious "gentleman's agreement" Landis did irreparable damage to his own legacy as well as that of Major League Baseball. The legacy of Landis "is always a complicated story" says MLB Historian John Thorn, but one that includes "documented racism."[29] The Baseball Almanac quotes Landis as once telling MLB players "… if you get into trouble, come to me. I'm your friend,"[30] but our great game's first Commissioner was no friend of integrity and honor when he permitted the gross discrimination against African American ballplayers to continue. It was only through the "courage, perseverance, and strength to overcome the oppressive racial segregation"[25] that we were fortunate enough to learn the names of Satchel Paige, Cool Papa Bell and Andrew "Rube" Foster (the "father of black baseball" and organizer of the Negro National League) as well as the other 32 Negro Leagues legends currently in the Major League Baseball Hall of Fame.

Josh Gibson, Buck Leonard, and "Smokey" Joe Williams are just some of those other African American superstars in Cooperstown. While Gibson and Leonard are relatively well known, Smokey Joe and too many others remain unknown, even to long-time fans of the game. Let us now remedy that injustice by recognizing and honoring each in the order of enshrinement: Satchel Paige, Josh Gibson, Buck Leonard, Monte Irvin, Cool Papa Bell, Judy Johnson, Oscar Charleston, Martin Dihigo, Pop Lloyd, Rube Foster, Ray Dandridge, Leon Day, Bill Foster, Willie Wells, "Bullet" Joe Rogan, "Smokey" Joe Williams, Turkey Stearnes, Hilton Smith, Ray Brown, Willard Brown, Andy Cooper, Frank Grant, Pete Hill, Biz Mackey, Effa Manley, Jose Mendez, Alex Pompez, Cumberland Posey, Louis Santop, Mule Suttles, Ben Taylor, Cristobal Torriente, Sol White, J.L. Wilkinson, and Jud Wilson.

Many of these names should be as recognizable as Aaron, Ruth, Mays, or Mantle. Why are they not? Landis shoulders much of the blame but in fairness it should be noted that greed also played a role in the hideous practice of segregation in baseball. According to documents in the Library of Congress, "… many owners of major league teams rented their stadiums to Negro League teams when their own teams were on the road. Team owners knew that if baseball were integrated, the Negro Leagues would probably not survive losing their best players to the majors," thus resulting in a significant loss of rental revenue.[31] Cool Papa Bell, who you may recall "smelled the rose of life" through baseball, once said "They say I was born too soon. I say the doors were opened too late."[26] Those doors remained closed for far too long because of repugnant intolerance and unbridled greed. Do those sins represent the greatest MLB scandal of all?

Fifth inning: was Marge Schott an anomaly or emblematic of cincinnati reds baseball culture?

Repugnance and intolerance in Major League Baseball did not end with Branch Rickey's humanity and Jackie Robinson's historic display of grace, courage, dignity, and sustained professional excellence. Instead, both tendencies became a visible hallmark of the Cincinnati Reds franchise for the duration of Marge Schott's ownership tenure from 1984 to 1999. Though the case can be made that Marge Schott was baseball's worst owner at the dawn of the new millennium, the insidious stain of discrimination within baseball's first professional franchise began long before Schott purchased her controlling interest and became the Reds' president and CEO.

According to Rick Swaine in *The Integration of Major League Baseball*, "In the early days of baseball's integration, Cincinnati… was a place that black players wished to avoid… Cincinnati fans were merciless in their abuse of black players at cozy Crosley Field where the crowd sat close enough to personally share their most intimate biases."[32] Warren Giles, who went on to serve as National League President for 18 years, ironically began his career as a major league baseball executive under Rickey in St. Louis, serving as general manager for multiple minor league teams. Following the 1936 season he was called up to the big leagues when he was tapped to replace Hall of Fame executive Larry MacPhail as the Cincinnati Reds' general manager. While Giles ultimately served a successful tenure as National League President, overseeing team moves to Atlanta, San Francisco and Los Angeles, expansion into New York, Houston, and Montreal, and the creation of divisional play and the National League Championship Series,[33] his time with the Reds was less distinguished.

Unfortunately, Giles never internalized Branch Rickey's inclination toward equality and integration while working for him in St. Louis. While representing the Reds organization in 1949, Giles echoed so many of the narrow minded of the time when he said "Recent performances by Negroes in the major leagues indicate that the outstanding Negro player can qualify for major league standard of play… although the ratio of Negro players to other players in the major leagues will probably be small for many years." Giles and the Reds "made no move to integrate during his tenure" in Cincinnati.[32]

While Giles was not considered a racist, his apathy toward equal treatment of black ballplayers represented a culture in Cincinnati baseball that resurfaced with Marge Schott's purchase of a controlling interest in the team in 1984. Schott's contributions and ultimate meaning to the city are complex, as she has been widely described as both generous and racist. Schott was a long-time philanthropist in Cincinnati, supporting causes as diverse as the city zoo, hospitals, and academic institutions. To this day, her Schott Foundation continues to support charitable programs that benefit the Cincinnati community. She supported fans by famously keeping ticket and concession prices low, with box seats the cheapest in baseball and hot dogs selling for a dollar long after every other owner had started gouging the hungry fan. Schott was a favorite of Cincinnati kids, too, often allowing groups of children to run around on the field after games. Yet there was another side to Marge Schott that awoke those ghosts of repugnant intolerance in baseball. It was that side that got Schott into trouble and ultimately suspended indefinitely by Major League Baseball's Executive Council.

In February of 1993 Schott was fined $25,000 and banned from day-to-day operations of the Reds for the 1993 season for allegedly making racial and ethnic slurs, including one as far back as 1987 on a conference call with other major league owners. Her actions were called "so embarrassing, so unacceptable and a threat to the interest of baseball" that owners had to do something. Again, however, the complexity of who Marge Schott was befuddled many observers. She immediately instituted an affirmative action plan for the Reds, hired Tony Perez (a Cuban and one of only two Hispanic managers in MLB), and employed minority-group members in the front office and throughout the minor leagues.[35]

Sports Illustrated's Rick Reilly had the opportunity to interview Schott in 1996 for the magazine. Reilly reported that over the years Schott had insulted homosexuals, blacks, Jews, and even other women, going on record preferring that the Reds "not hire women of child-bearing age." She took a compassionate view of Hitler, saying "He was O.K. at the beginning... He just went too far". She kept a Nazi armband in a bureau drawer in her home and told ABC's Diane Sawyer that the armband "is not a symbol of evil to me." Reilly relayed stories of Schott negatively stereotyping Asian-Americans and the Japanese as well, using a cartoonish accent to ridicule.[36] Schott's comments about Hitler were apparently the last straw. Major League Baseball forced Schott to step aside on June 12th, 1996. The controversial owner was compelled to surrender day-to-day control of the team through the 1998 season, the second time in 3 years that Schott had to relinquish leadership of the Reds rather than challenge a suspension in court. Her tendency to bring "disrepute and embarrassment"[37] to baseball with her repeated use of racial and ethnic slurs continued until 1999 when, facing a possible third suspension, she sold her controlling interest in the Reds to a Cincinnati businessman.

Like many other baseball organizations in the middle of the 20th century, the Cincinnati Reds displayed a gross mistreatment of black ballplayers. It was not until 1951 when Giles left the Reds organization for the National League Presidency that Gabe Paul assumed duties as general manager and showed an interest in integrating the Reds. However, that same hostility toward black baseball players seemed to be resurrected with the arrival of Marge Schott, only she never seemed to know it. Baseball's second female owner, who funded children's hospitals, kept prices low for fans, and regularly showed her love for young fans, when asked by MLB's Executive Council to relinquish her duties (running the Reds) or face indefinite suspension for her repeatedly racist comments, told the council "I don't see that I said anything wrong".[38] Cincinnati Enquirer sports columnist Paul Daugherty may have put it best... "The true measure of a person is whether he or she leaves a place better than he or she found it. Is Cincinnati better for having raised and rooted for Marge Schott?... Good Marge competed with Bad Marge daily."[34]

In August of 2020, longtime Reds announcer Thom Brennaman unwittingly contributed to the possible perception that the intolerance demonstrated within the Reds organization throughout the years is not just about Marge Schott but is part of a larger organizational culture. Brennaman made a homophobic slur on a hot microphone during the first game of a double-header against the Kansas City Royals. Brennaman apparently thought the telecast was still on commercial break when he inadvertently made the on-air anti-gay insult. He followed his actions with the standard post-game mea culpa…"I can't begin to tell you how sorry I am. That is not who I am, it never has been."[47] That may or may not be who Thom Brennaman is; only he knows for sure. What the Reds organization seems to know, or at least wants you to know, is that the actions of Brennaman (just like the actions of Giles, whose apathetic attitude toward the equal treatment of black ball-players is also a stain on Cincinnati) do not represent Reds baseball. "The Cincinnati Reds organization is devastated by the horrific, homophobic remark made this evening by Reds broadcaster Thom Brennaman… In no way does this incident represent our players, coaches, organization or our fans… The Reds embrace a zero-tolerance policy for bias and discrimination of any kind."[48] So the question remains: Was Marge Schott an anomaly or emblematic of an intolerant Cincinnati Reds baseball culture?

Sixth inning: was the 1919 world series fix a foreshadowing of the future?

The Cincinnati Reds were deeply involved in another of MLB's all-time great controversies, only it was the Chicago White Sox organization which was at fault and was ultimately rebranded the Black Sox for what is arguably the most notorious gambling scandal in U.S. sports history. On October 1, 1919, the two teams played Game 1 of the World Series. Eight days later the Reds were crowned champions, having vanquished the Sox five games to three (from 1919 to 1921 a 5 of 9 game series was played), but was the victory legitimate? Though no players were ever convicted of any charges, Shoeless Joe Jackson, Eddie "Knuckles" Cicotte, and six of their teammates will forever be known as "Eight Men Out."

The sequence of events is straightforward. In his book *Burying the Black Sox*, baseball historian Gene Carney lays out the succinct timeline below,[39] starting with Joe Jackson's ground out to second base ending the Reds' 10-5 series clinching victory on October 9th.[40]

October 1919—Reds win the 1919 World Series 5 games to 3; Reporter Hugh Fullerton, quoting Sox owner Charles Comiskey the following day, writes that seven Sox players will not return.

December 1919—Fullerton writes a series of articles calling on baseball to investigate fix rumors.

Spring 1920—Lee Magee, who played multiple positions for multiple teams, sues baseball, claiming he was "blacklisted" after being suspected of crooked play. Magee loses a trial in June.

August 1920—The Cubs and Phillies play a game of little significance to the pennant race but rumors that a fix was in cause a stir.

September 1920—A grand jury in Chicago is directed to investigate the rumors about the Cubs-Phillies game and the gambling problems of Major League Baseball. Later that month, a Philadelphia reporter breaks a story about the 1919 World Series being fixed. White Sox pitcher Cicotte then confirms to the grand jury that ballplayers and gamblers had plotted to toss the World Series. Shoeless Joe and Sox pitcher Lefty Williams also testify to the grand jury. Eight players are indicted and immediately suspended (Cicotte, Jackson, Williams, 1B Chick Gandil, CF Oscar Felsch, SS Swede Risberg, 3B George Weaver, and infielder Fred McMullin).

November 1920—Kenesaw Mountain Landis is named baseball's first Commissioner.

December 1920—Evidence from the grand jury hearings vanishes.

March 1921—New indictments are issued for the ballplayers as well as ten gamblers.

June 1921—The Black Sox trial begins in Chicago.

August 1921—Neither the players nor the gamblers are found guilty; Landis permanently bans the players from baseball.

Some argue that the games "thrown" by the White Sox players were not the real scandal. Instead, like the old political adage goes, it is not the crime but the cover-up. No serious investigation was ever conducted. The roles of others, from ownership to gamblers to everyone else making money from the enterprise of baseball were simply ignored. Carney writes, "… we see corporate America reflected in the White Sox and Major League Baseball, silencing whistle-blowers, keeping real investigators off the trail, controlling the damage and the spin, and finally, when the lid is blown, keeping the focus on eight employees, as if they alone had 'guilty knowledge'".[39] Gambling in baseball and the subsequent efforts to conceal it were no secret early in the 20th century. "The players knew it was going on, and the owners knew it was going on. But more important, the players knew that the owners knew—and they knew that the owners were doing nothing about it."[41]

The infusion of legalized gambling into modern American sport may ultimately reveal that betting and the Black Sox scandal were simply a foretelling of more serious problems to come for Major League Baseball. Wagering on baseball was always an unofficial national pastime indulged by all classes of American society, but as baseball became more fully organized (and profitable for many), professional gamblers moved in and the game's integrity suffered. In *Eight Men Out*, Eliot Asinof shares unbelievable tales of how gambling impacted the game. "An outfielder, settling under a crucial fly ball, would find himself stoned by a nearby spectator, who might win a few hundred dollars if the ball was dropped. On one occasion, a gambler actually ran out on the field and tackled a ballplayer. On another, a marksman prevented a fielder from chasing a long hit by peppering the ground around his feet with bullets."[42] As time went on Major League Baseball ostensibly worked harder to protect the integrity of the game from gambling… our First Inning discussion on Pete Rose's punishment representing a good example. However, when they say it is not about the money, rest assured that it is about the money. The Brotherhood of Professional Baseball Players knew this well. Launched in 1885, the Brotherhood represented the first serious effort to organize a labor union consisting of baseball players.[43] The organization's 19th century statement of purpose included this prescient take on where greed would ultimately lead Major League Baseball:

"There was a time when the League stood for integrity and fair dealing. Today it stands for dollars and cents. Once it looked to the elevation of the game and an honest exhibition of the sport; today its eyes are on the turnstiles. Men have come into the business for no other motive than to exploit it for every dollar in sight."[42]

Shoeless Joe and Lefty Williams had figured this out. Carney writes that on the day that the 1920 World Series opened, Jackson and Williams were in Greenville, South Carolina meeting with a local attorney. Under indictment, they realized that Comiskey was less concerned with their interests than in protecting his own investment, even though he was widely suspected of having known of the Black Sox intentions. Quoted in the Chicago Times on October 6, 1920, Shoeless Joe Jackson and Lefty Williams had "no further comment to make other than the statement that if the investigators probe thoroughly they may find men higher up in baseball at the bottom of the (Black Sox) scandal".[39]

The Brotherhood of Professional Baseball Players knew in 1885 that "Men have come into the business for no other motive than to exploit it for every dollar in sight." Indicted ballplayers in 1919 knew that owners cared less about their employees and more about their own bottom line. Today, approximately 50% of US residents are projected to live in a state that has launched legal sports betting by the year 2024. As America's pastime, baseball is one of the most popular sports on which to bet. Sportsbooks around the world cater to gamblers wagering billions of dollars on Major League Baseball, resulting in millions of dollars in revenue.[44] The game has not let the (heretofore sordid) opportunity to align itself with gambling to pass. An early investor in DraftKings, a company marketing itself as "opportunity knocking… the prime rib for hungry American sports dogs" and a place to "Dig in to all your favorite sports wagers,"[45] Major League Baseball has gone all in on the very activity that got the Hit King banned for life. Dustin Gouker, a longtime sports journalist, recently reported that "DraftKings and

Major League Baseball made their relationship a little tighter, as the sports betting company became an official league partner for gaming (betting)". Surprised? Do not be. The deal with DraftKings was MLB's third agreement with the sports betting industry after earlier partnership deals with MGM Resorts International[46] and FanDuel.

Professional sports leagues in the United States consistently insist that everything they do is "about the fans." Really? Many of us can recall afternoon World Series games from our childhood but in 2020 we can barely remain awake to get through the fifth inning of the Fall Classic because all World Series games now start well after 8:00 pm Eastern Standard Time. Are late night starts 'for the fans' or a function of television revenue? Likewise, is promulgating gambling, an endeavor that mathematics makes clear is a losing proposition, 'for the fans' or simply another source of profit for MLB? How about $9.00 hot dogs, $13.00 beers, or $45.00 parking… 'for the fans' or maximization of income, working families be damned? The 1919 Black Sox scandal is another stain on Major League Baseball, make no mistake. But with the MLB embracing gambling and all the issues that come with it, the real question is were the actions of Shoeless Joe and the boys a foreshadowing of the future of America's pastime?

Seventh inning: do the St. Louis Cardinals bear some responsibility for baseball's sign stealing scandal?

Buzzers, garbage cans, and illicit video room reviews… those stories dominated baseball in 2019 as much as anything the game's young stars or aging heroes did on the field. We will get to the Astros' sign stealing scandal in the Eighth Inning but first we must address what can be considered a prelude to that forbidden scheme—the St. Louis Cardinals' 2015 database hacking crime. Who was the victim of the Redbirds' high-tech underhandedness? None other than the (now vilified) Houston Astros.

Only 4 years before Altuve, Bregman, and Hinch collectively became public enemy #1, the Astros were seen as innocent victims of a complex espionage attack against their scouting system, draft and trade discussions, and other proprietary data. Christopher Correa, former Scouting Director for the St. Louis Cardinals, pleaded guilty in January of 2016 to five counts of unauthorized access of computers belonging to employees of the Houston Astros beginning in 2013. He was sentenced to nearly 4 years in prison and ordered to pay $279,038 in July 2016. According to the New York Times, Correa told the judge "I violated my values, and it was wrong," and that he was "overwhelmed with remorse and regret for my actions."[49]

It all started when Cardinals owner Bill DeWitt hired Jeff Luhnow as Vice President of Baseball Development in 2003. Luhnow was a baseball outsider who was unexpectedly offered an interview with DeWitt after DeWitt's son-in-law told the owner about a business contact (Luhnow) possessing strong managerial skills. "Came out of nowhere for me," Luhnow said in 2014, "but when the owner of a baseball team wants to talk to you, you talk to him."[50] Luhnow immediately created the Cardinals' analytics department, and in 2009 hired a University of Michigan doctoral student who had completed a couple of freelance projects for the organization. Those projects involved designing software that could scrape play-by-play data from the websites of college teams across the country, even those in Division II and III, and centralize that data, giving the Cardinals' analytics department a resource that most other MLB teams did not have.[51] That young software designer was Christopher Correa.

Correa immediately had an impact working on the 2009 draft under Luhnow; as Ben Reiter reported in Sports Illustrated, Correa's work unearthed Slippery Rock's Matt Adams in the 23rd round of that year's draft, a proverbial diamond in the rough. Adams has gone on to have a long and successful career, highlighted by his 2019 World Series championship with the Washington Nationals. Other solid Luhnow-Correa picks in that 2009 draft, which is widely considered one of the best ever, include Joe Kelly, Shelby Miller, Trevor Rosenthal, and Matt Carpenter, all of whom made up the core of the 2013 National League pennant-winning Cardinals team with Matt Adams. It was clear that Luhnow and Correa had successfully introduced complex analytics to one of baseball's oldest and most prosperous franchises, and the two had a bright future together in St. Louis.

That future faded when Luhnow left the Cardinals organization for the General Manager position with the Houston Astros in December of 2011. Luhnow ultimately took some of his Cardinals colleagues with him to Houston, including Sig Mejdal, a former NASA scientist who worked with Luhnow in St. Louis to build the Cardinals' internal database for baseball operations. While that database, called Redbird, proved to be a valuable resource for the Cardinals, Reiter reported that many inside the St. Louis front office "saw (Luhnow) as a know-it-all outsider who, despite his complete lack of a baseball background, believed he knew how to do things better than they did." Reiter's reporting offers a plausible motive for the felonious activity that ultimately transpired 24 months later—the only thing more maddening for certain Cardinals officials than working with Luhnow was working against him.[50]

Luhnow, Mejdal, and others ultimately created a proprietary internal database in Houston called Ground Control. While doing a Sports Illustrated story on the Astros rebuilding effort in 2014, Reiter reported that Ground Control was an impressive and multifaceted database. He described it as containing "scouting reports (complete with on-demand video), statistics, injury histories, and projections for every professional player and amateur prospect on the pro radar" and that it "suggests the optimal defensive shift to be deployed against every opposing batter". Finally, Reiter reported, Ground Control also allowed team executives to view the locations and schedules of every one of their scouts and to figure out how to deploy them most efficiently.[50] Luhnow and Mejdal had created the gold standard for an analytical database in baseball, and their old friends in St. Louis had taken notice.

In 2015 an investigation by the F.B.I. and Justice Department prosecutors addressed accusations of hacking into Ground Control to steal the contents of the Astros database on professional and amateur ballplayers. The investigation began when information on internal Astros trade talks was anonymously leaked to the now defunct website, Anonbin, and ultimately appeared on the website Deadspin in June 2014. After immediately confirming that the information was not leaked by anyone within the organization, the Astros worked with MLB and federal authorities to uncover "evidence that Cardinals employees broke into a network of the Astros that housed special databases (Ground Control) the team had built."[52] Ultimately the Cardinals and Major League Baseball were served subpoenas by the F.B.I.'s Houston field office for relevant electronic correspondence. The Assistant U.S. Attorney handling the case in Houston was Michael Chu, an expert in computer hacking and intellectual property. Chu, working with the F.B.I., traced the breach back to a house in Jupiter, Florida, the city in which the Cardinals hold spring training. Cardinals employees had occupied the house that spring.[53]

The investigation was swift and ultimately revealed that the Astros' database was accessed at least 48 times. Correa entered a guilty plea on January 8, 2016, only 7 months after he ran his first and only draft for the Cardinals (he picked outfielder Harrison Bader, pitcher Jordan Hicks, and shortstop Paul DeJong, all of whom were on the Cardinals' roster within 4 years). Correa agreed to plead guilty for two reasons: "One, I was guilty. Two, I wanted to accept responsibility... so I could move on with my life."[51] A month after that 2015 draft Correa was fired by the Cardinals, and a year later he returned to the courtroom for sentencing. In January of 2017, Commissioner Rob Manfred placed Correa on MLB's permanently ineligible list.

Christopher Correa had risen quickly but fallen even faster. From doctoral student to software analytics designer to Scouting Director for one of baseball's most storied franchises, Correa entered the federal prison camp in Cumberland, Maryland on August 30, 2016. Only 14 months had passed between the height of Correa's professional accomplishments (leading the June 2015 MLB draft for the Cardinals) and assuming his new identity (Federal Inmate #04550-479). Perhaps adding insult to injury, half of Correa's wages in the prison camp, which ranged between

12 and 40 cents per hour, were garnished by the Astros at the order of the judge in the case.[51]

The complexity of Correa's case and the multiple personnel angles involved make it seem difficult to comprehend, but that is not necessarily so. In fact, rather than complex some found it amusing. Correa simply used Mejdal's old Cardinals password to access the Astros' system, a fact that tickled venerable MLB veteran Torii Hunter. "It's funny to me," Hunter quipped at the time, "it's like you have a weakness and your firewall ain't good, it's like shoot, we're going to take your information. Houston just hung a curveball and now the Cardinals are banging."[54] Mejdal, the former NASA scientist and Cardinals analyst is the one who hung that "curveball" when he left St. Louis to (ironically) take the position of Director of Decision Sciences with Houston. Mejdal made a scientifically bad decision in leaving his Houston database password essentially the same as his old password in St. Louis. Alongside Correa, Mejdal should have done more to prevent the sequence of events that led to the damage done to the Houston organization. The Astros, for their part, greatly lamented that they were the only victims in the case, suffering grave losses of proprietary information that provided others a competitive advantage. Surely the Astros would have to exact revenge and restore that competitive balance somehow. So that brings us back to the original question: Do the St. Louis Cardinals bear some responsibility for baseball's sign stealing scandal?

Eighth inning: was the Astros' sign stealing chicanery the worst scandal since segregation?

Christopher Correa did not believe he had committed a crime when he repeatedly accessed the Houston Astros' database, Ground Control, instead believing what he had done was simply high-tech sign stealing. "It was all in the context of a game, to me," he said shortly after being transferred from the federal prison camp to home confinement. "When a pitcher throws at a batter's chest, nobody runs to the local authorities and tries to file an assault charge... If another team does something wrong, you retaliate."[51] Perhaps Correa's philosophy of retaliation is what motivated the Astros to engage in what some believe is the worst scandal in MLB since segregation. However, as Paul Harvey used to remind us, there is often more to the story.

On the eve of the 2017 playoffs, Major League Baseball initiated an investigation of sign stealing. Quick, name the team implicated in that illicit scheme. If you said the Houston Astros, you are not alone in that mistake. Late in the 2017 season the Boston Red Sox came under scrutiny for using an Apple watch to aid in stealing signs against the New York

Yankees and other MLB teams. Staff personnel with the responsibility of monitoring instant replay video would pass the signs to trainers via an Apple Watch. The trainer would then communicate with Red Sox players in the dugout, who in turn would relay what pitch was coming to their teammate in the batter's box. In one piece of video evidence, Dustin Pedroia is seen receiving information from a trainer and passing it to outfielder Chris Young.[55] The Red Sox were ultimately fined by Commissioner Rob Manfred, who was quoted as saying "We're 100% comfortable that it is not an ongoing issue."[56] Unfortunately, Manfred got that wrong, as sign stealing had a robust past and an even more prosperous future.

Today, no one seems to remember the Red Sox and their trickery (nor do most fans remember that the Yankees were also fined in 2017 for improper use of a dugout telephone to assist with sign stealing). Likewise, very few cite the 1951 Giants' purported use of a telescope to facilitate Bobby Thomson's "Shot Heard Round the World" off Dodgers' pitcher Ralph Branca. Al Gettel, a pitcher on that Giants team that won baseball's most famous pennant race, was unambiguous about the club's sign stealing. "Every hitter knew what was coming..." Gettel said, "made a big difference."[57] In 1973, Rangers manager Whitey Herzog filed a complaint that Bernie Brewer, the Milwaukee Brewers mascot clad in Bavarian clothing, relayed signs to Milwaukee hitters by clapping with his white gloves (or not) while sitting in the outfield bleachers at County Stadium.[58] Accusations against the Mets in 1997 that Bobby Valentine (ironically, Branca's son-in-law) planted cameras in Shea Stadium or charges against the Indians that they did the same at Jacobs Field in 1999 are rarely heard today. More recently, multiple teams lodged charges against Philadelphia in 2011 alleging that the Phillies used binoculars and otherwise violated league rules to steal signs. Do you remember allegations by multiple teams against the Phillies for sign stealing? How is it that only the Astros are considered villains in 2020, suffering public scorn and repeated beanings? Teams sent the Astros an immediate message following former Astros pitcher Mike Fiers's 2019 public allegations that the Astros cheated by engaging in sign stealing during their World Series winning 2017 season. Seven different Astros players were hit by a pitch in the first 5 games of the 2020 spring training season,[59] and the contempt (and beanballs) continued throughout the pandemic-shortened regular season until the Astros were eliminated in the 2020 American League Championship Series by the Tampa Bay Rays.

Baseball's Commissioner was "100 percent comfortable" in the fall of 2017 that sign stealing was not an ongoing issue, but in fact the practice was common around the game and had been for more than 70 years. While baseball had heretofore treated what Tom Koch-Weser, Astros Director of Advance Information, called "'the dark arts' of cheating with a wink and a nod,"[58] the World Series winning Astros became the poster child for ethical indignation after their sign stealing charade became widely known. They immediately became the subject of vitriol and scorn throughout MLB clubhouses and in the stands. Yet how did their actions, using video and garbage can covers (the allegation of the use of buzzers was debunked by Manfred) to steal signs, differ from the actions of Bobby Thomson's Giants team, Bernie Brewer, or Bobby Valentine? The short answer is there was no difference, but MLB nonetheless imposed the most severe punishment on a baseball franchise in memory: 1-year suspension of Manager A.J. Hinch and General Manager Jeff Luhnow (yes, that Jeff Luhnow), $5 million fine, and loss of first and second round picks in the 2020 and 2021 Major League Baseball draft.

To be sure, long before the sanctions were imposed on the Astros, Commissioner Manfred was unambiguous about the future. In a statement on September 15, 2017 announcing his relatively modest fines of the Red Sox and Yankees for sign stealing, Manfred said "All 30 clubs have been notified that future violations of this type will be subject to more serious sanctions, including the possible loss of draft picks". Inexplicably, subsequent investigations revealed that the Astros used their video feed and a system of banging on a trash can to alert their hitters of the pitch that was coming just 6 days later in a game against the White Sox. High-tech sign stealing continued, and not just in Houston. After spending "the entire 2017 postseason running back and forth chasing down allegations about who was stealing signs… everybody was charging everybody with doing it", MLB reiterated its stance. In a March 27, 2018 memorandum to all clubs, Manfred again warned, "Electronic equipment, including game feeds in the club replay room and/or video room, may never be used during a game for the purpose of stealing the opposing team's signs."[58] The Commissioner cautioned all teams that future violation of these rules would result in severe punishment, and that is exactly what was imposed on the Houston Astros.

In 1926, Ty Cobb wrote in a newspaper column that "In the minds of the public, there seems to be an impression that sign stealing is illegal… It is not so regarded by ball players. If a player is smart enough to solve the opposing system of signals, he is given due credit". However, the Georgia Peach drew the line at the use of anything more than one's intellect. "There is a form of sign stealing which is reprehensible and should be so regarded… the use of outside devices is against all the laws of baseball and the playing rules."[60] The public obviously considers (electronic) sign stealing to be illegal, as evidenced by the continuing contempt heaped upon Jose Altuve, Alex Bregman, and the rest of the Houston organization after their use of video recording to gain a competitive advantage. But remember what Christopher Correa said after his conviction for hacking the Astros database in 2016: "If another team does something wrong, you retaliate." While the Houston Astros issued a statement at the conclusion of the hacking scandal that indicated they were "looking forward to focusing our attention on the 2017 season and the game of baseball,"[61] perhaps they were actually more focused on the material harm they endured through the database hacking and on retaliation through electronic sign stealing. So, I ask you again: Was the Astros' sign stealing chicanery the worst scandal since segregation, or was it simple retaliation or even just standard long-standing sign stealing practice throughout baseball?

Ninth inning: can baseball ultimately reclaim its soul?

If another team does something wrong, you retaliate. Christopher Correa's inclination to rationalize his hacking actions against the Astros, and the Astros' subsequent use of electronic sign stealing to regain a competitive edge are but two instances of baseball's descent into what Koch-Weser called baseball's "dark arts."[62] Recent examples of unethical practices include both simple and complex schemes to subvert the rules:

- The Angels' clubhouse attendant accused of "concocting and distributing a signature mix of pine tar and rosin designed to improve pitchers' grips on the baseball"[63]
- Pitchers across MLB increasingly using foreign substances on the baseball[64]
- Padres GM A.J. Preller hiding medical records from trade partners[64]
- Braves GM John Coppolella violating international signing rules[64]
- Christopher Correa's database hacking scheme[64]
- The Yankees', Red Sox', and Astros' electronic sign stealing revelations[65]

Eighteenth-century philosopher Immanuel Kant posited that the supreme principle of morality is a standard of rationality that he dubbed the "categorical imperative" (CI). Kant characterized the CI as an objective, rationally necessary, and unconditional principle that we must always follow despite any natural desires we may have to the contrary.[66] Kant suggested that the CI is "universal and impartial—universal because all people, in virtue of being rational, would act the same way, and impartial because their actions are not guided by their own biases (or circumstances), but because they respect the dignity and autonomy of every human being and do not put their own personal ambitions above the respect others deserve."[67] In other words,

Kant told us, Correa's retaliation philosophy misses the mark on how people ought to behave—if an action is wrong, it is always wrong, regardless of the situation.

Baseball has been getting a lot wrong lately and the results are glaring. In addition to what seems like constant subversion of the rules of the game, myriad changes arising from the Commissioner's office alongside how players and managers execute on the field are also wreaking havoc on both the essence of baseball and its popularity with fans. New York sportswriter Phil Mushnick declares, "If you inspect the significant changes to MLB in the Bud Selig and now Rob Manfred eras, it's difficult to find one designed to benefit the sport rather than the teams' bottom lines—and at the game's and its fans' expense."[1] On field changes are not much better for the fan. Games regularly run well over 3 hours with a dreadfully slow pace of play—"one ball put into play every 3 minutes, 42 seconds."[62] How is that possible? Fundamental baseball has yielded to analytics and exit velocity. From 1980 to 2016, home runs were up 32%, strikeouts increased 67%, while balls put in play descended to a record low.[17] In 2018, for the first time in MLB history, there were more strikeouts than hits—want more? As recently as 2008, there were 11,000 more hits than strikeouts.[68,69] In just 10 years, the hits to strikeouts ratio inverted drastically. Our beloved game has survived scandal and controversy for more than 150 years since the founding of baseball's first professional organization, the Cincinnati Red Stockings in 1869. However, recent decisions both on and off the field have put the game's future at stake.

Shoeless Joe and Pete Rose and gambling. Jackie Robinson and Cool Papa Bell and segregation. Kenesaw Mountain Landis and Marge Schott and racism. Christopher Correa and the World Series Champion, Astros and electronic espionage. Baseball's history is littered with scandal and controversy, but Sports Illustrated's Tom Verducci argues that we are now at a tipping point. "Baseball is in a fight to reclaim its soul. That soul of the game must be found in its aspirational value: players of all sizes playing a simple kids' game." MLB Players Association Union Chief Tony Clark seems to agree. Clark contends that the way in which MLB and the players handle the immediate future "will significantly affect what our game looks like for the next several decades."[62]

In 1956 Robert Frost declared that "Baseball is the fate of us all... I am never more at home in America than at a baseball game."[70] There is really no need to repeat the Ninth Inning question. Baseball can... baseball must... baseball will... reclaim its soul and bring all fans home. I suggest MLB starts by banning the defensive shift and nullifying Ryan for Fregosi...

References

1. Mushnick, Phil. *Eyes on the Ball*, New York Post, February 14, 2020.
2. Serena, Katie. *These '70s Yankees Pitchers Wanted Each Other's Wives – So They Made The Trade Of The Century*, https://allthatsinteresting.com/mike-kekich-fritz-peterson, April 9, 2018.
3. Gentile, James. *The Best Bunter of All-Time*, https://www.beyondtheboxscore.com/2012/12/10/3748738/best-bunter-all-time-career-bunt-hits-bases-empty-mlb, December 10, 2012.
4. Verducci, Tom. *Pete Rose on his Hall of Fame Hopes, Steroids in Baseball, and Much More*, https://www.si.com/mlb/2015/07/14/pete-rose-interview-all-star-game, July 14, 2015.
5. Associated Press as reported by the Los Angeles Times. *Pete Rose Banned for Life: Giamatti Says He Bet on Games*, https://www.latimes.com/archives/la-xpm-1989-08-24-mn-1531-story.html, August 24, 1989.
6. Ducey, Kenny. *Pete Rose to Remain Banned from Baseball*, https://www.si.com/mlb/2015/12/14/pete-rose-rob-manfred-baseball-ban-hall-of-fame, December 14, 2015.
7. Gartland, Dan. *Commissioner Rob Manfred will speak with Pete Rose about Reinstatement*, https://www.si.com/mlb/2015/03/16/commissioner-rob-man-fred-pete-rose-reinstatement, March 16, 2015.
8. Verducci, Tom. *Why Rob Manfred Shut the Door on Pete Rose's Reinstatement Hopes*, https://www.si.com/mlb/2015/12/14/pete-rose-reinstatement-lifetime-ban-rob-manfred, December 14, 2015.
9. Sports Illustrated Wire. *Pete Rose's Investigator Advocates Maintaining Lifetime Ban*, https://www.si.com/mlb/2015/03/22/pete-rose-investigator, March 22, 2015.
10. Kennedy, Kostya. *Report that Pete Rose Bet on Games as a Player Should Surprise No One*, https://www.si.com/mlb/2015/06/22/pete-rose-betting-games-player-gambling-cincinnati-reds, June 22, 2015.
11. Sullivan, Tim. *Truth is, Pete Rose Can't Be Trusted*, https://www.courier-journal.com/story/sports/tim-sullivan/2015/06/22/truth-pete-rose-trusted/29133279/, June 23, 2015.
12. Schmidt, Michael. *Dear Pete Rose: It's Still a No. Sincerely, Baseball* https://www.nytimes.com/2015/12/15/sports/baseball/pete-rose-ban-mlb-commis-sioner-rob-manfred.html?smid=tw-share, December 14, 2015.
13. Sports Illustrated Wire. *Ex-Cincinnati Red Barry Larkin: Pete Rose Belongs in the Hall of Fame*, https://www.si.com/mlb/2015/04/25/barry-larkin-pete-rose-hall-fame, April 24 2015.
14. Sports Illustrated Wire. *Cal Ripken Jr. on SI Now: Pete Rose Belongs in the Hall of Fame*, https://www.si.com/mlb/2015/04/09/pete-rose-hall-fame-reds-cal-ripken-jr-si-now, April 9, 2015.
15. Dey, Who. *4,256 Reasons Why Pete Rose Belongs in the Hall of Fame*, https://bleacherreport.com/articles/6876-4256-reasons-why-pete-rose-belongs-in-the-hall-of-fame, January 17, 2008.
16. Genco, Raymond and Rosenbaum, Mark. *Letter to the National Baseball Hall of Fame President*, September 27, 2016.

17. Kennedy, Merrit. *Major League Baseball Poised to Change Intentional Walk Rule*, https://www.npr.org/sections/thetwo-way/2017/02/23/516878572/major-league-baseball-poised-to-change-intentional-walk-rule, February 23, 2017.

18. Baseball Reference.com. *All-Century Team*, https://www.baseball-reference.com/bullpen/All-Century_Team, 1999.

19. Baseball Reference.com. *Jimmie Foxx*, https://www.baseball-reference.com/players/f/foxxji01.shtml, 2020.

20. Baseball Reference.com. *Overall Baseball Leaders and Baseball Records*, https://www.baseball-reference.com/leaders/, 2019.

21. Pepe, Phil. *The Original Wife Swap: Yankees Pitchers Fritz Peterson and Mike Kekich Trade Wives*, New York Daily News, https://www.nydailynews.com/sports/baseball/yankees/original-wife-swap-yankee-pitchers-trade-lives-article-1.2138703, March 6, 1973.

22. Capozzi, Joe. *Ex-Yankee Fritz Peterson has No Regrets 40 Years After Wife Swap*, Palm Beach Post, https://en.wikipedia.org/wiki/Fritz_Peterson, January 26, 2013.

23. Abraham, Tamara. *We Didn't Just Swap Wives, We Swapped Lives: How Yankees Stars Traded Families in Scandal that Rocked Seventies Baseball*, https://www.dailymail.co.uk/femail/article-2006350/We-didnt-just-swap-wives-swapped-lives-How-Yankees-stars-traded-families-scandal-rocked-Seventies-baseball.html, June 21, 2011.

24. National Baseball Hall of Fame. *Cool Papa Bell*, https://baseballhall.org/hall-of-famers/bell-cool-papa, 1 September 2020.

25. U-S-History.com. *Negro Leagues*, https://u-s-history.com/pages/h2079.html, August 30, 2020.

26. Web Circle. *Baseball Quote of the Day*, http://quote.webcircle.com/cgi-bin/search.cgi?team=Grays, August 30, 2020.

27. Rickey, Branch. *One Hundred Percent Wrong Club* (Speech, Atlanta, Georgia), https://www.loc.gov/collections/jackie-robinson-baseball/articles-and-essays/baseball-the-color-line-and-jackie-robinson/one-hundred-percent-wrong-club-speech, January 20, 1956.

28. U-S-History.com. *Kenesaw Mountain Landis*, https://u-s-history.com/pages/h2074.html, August 30, 2020.

29. Gaydos, Ryan. *House Dems Call on MLB to Remove Kenesaw Mountain Landis' Name from MVP Awards*, https://www.foxnews.com/sports/house-dems-call-mlb-remove-kenesaw-mountain-landis-name-mvp-awards, August 6, 2020.

30. Baseball Almanac. *Commissioner Kenesaw Landis Biography*, https://www.baseball-almanac.com/articles/kenesaw_landis_biography.shtml, August 30, 2020.

31. Library of Congress. *Breaking the Color Line: 1940-1946*, https://www.loc.gov/collections/jackie-robinson-baseball/articles-and-essays/baseball-the-color-line-and-jackie-robinson/1940-to-1946/#:~:text=By%20the%201940s%2C%20organized%20baseball,racially%20segregated%20for%20many%20years.&text=In%20addition%20to%20racial%20intolerance,teams%20were%20on%20the%20road, August 30, 2020.

32. Swaine, Rick. *The Integration of Major League Baseball*, McFarland & Company, Inc., Jefferson, North Carolina, 2009.

33. National Baseball Hall of Fame. *Warren Giles*, https://baseballhall.org/hall-of-famers/giles-warren, 24 September 2020.

34. Goldman, Tom. *She was Generous. She was Also Racist. Should This Ballpark Carry Her Name?*, https://www.npr.org/2020/06/10/873511957/a-stadium-name-once-meant-to-honor-now-offends, June 10, 2020.

35. Smith, Claire. *Baseball; Owners' Ruling on Schott is Assailed and Lauded*, https://www.nytimes.com/1993/02/04/sports/baseball-owners-ruling-on-schott-is-assailed-and-lauded.html?searchResultPosition=10, February 4, 1993.

36. Reilly, Rick. *Heaven Help Marge Schott*, https://www.si.com/mlb/2014/11/19/heaven-help-marge-schott-rick-reilly-si-60, May 20, 1996.

37. Associated Press. *Baseball Slips Muzzle on Marge Through '98*, https://www.spokesman.com/stories/1996/jun/13/baseball-slips-muzzle-on-marge-through-98/, June 13, 1996.

38. McCoy, Hal. *Schott Faces Indefinite Suspension, Sources Indicate Announcement Expected Today*, https://www.spokesman.com/stories/1996/jun/12/schott-faces-indefinite-suspension-sources/, June 12, 1996.

39. Carney, Gene. *Burying the Black Sox: How Baseball's Cover-Up of the 1919 World Series Fix Almost Succeeded*, Potomac Books Inc., Dulles, Virginia, 2006.

40. Retrosheet Boxscore. *World Series Game 8, October 9, 1919*, https://www.retrosheet.org/boxesetc/1919/B10090CHA1919.htm.

41. Gardner, Paul. *Nice Guys Finish Last: Sport and American Life*, Universe Books, 1975.

42. Asinof, Eliot. *Eight Men Out*, Holt, Rinehart and Winston, New York, 1963.

43. Baseball Reference.com. *Brotherhood of Professional Baseball Players*, https://www.baseball-reference.com/bullpen/Brotherhood_of_Professional_Baseball_Players, 2020.

44. Legal Sports Betting. *How Much Money Do Americans Bet on Sports?*, https://www.legalsportsbetting.com/how-much-money-do-americans-bet-on-sports/, September 22, 2020.

45. DraftKings.com. *DraftKings Begins Trading on the NASDAQ Stock Exchange*, https://www.draftkings.com/about/, October 4, 2020.

46. Gouker, Dustin. *DraftKings Does Deal with MLB to Become an Authorized Gaming Operator*, https://www.legalsportsreport.com/34611/draftkings-mlb-sports-betting-deal/, October 4, 2020.

47. Canova, Daniel. *Cincinnati Reds Suspend Announcer Thom Brennaman After 'Horrific' Anti-gay Slur on Air*, https://www.foxnews.com/sports/cincinnati-reds-suspend-announcer-thom-brennaman-after-horrific-anti-gay-slur-on-air, August 20, 2020.

48. Planalp, Brian. *Reds Suspend Brennaman After 'Horrific" On-Air Remark*, https://www.fox19.com/2020/08/19/reds-issue-statement-brennamans-horrific-remark-air/, August 20, 2020.

49. The Associated Press. *Christopher Correa, a Former Cardinals Executive, Sentenced to Four Years for Hacking Astros*, https://www.nytimes.com/2016/07/19/sports/baseball/christopher-correa-a-former-cardinals-executive-sentenced-to-four-years-for-hacking-astros-database.html#:~:text=HOUSTON%20%E2%80%94%20A%20federal%20judge%20sentenced,two%20Major%20League%20Baseball%20clubs, July 18, 2016.

50. Reiter, Ben. *Cardinals' Hacking into Astros' Database is Worse than Deflategate*, https://www.si.com/mlb/2015/06/16/astros-cardinals-fbi-hacking-ground-control, June 16, 2015.

51. Reiter, Ben. *What Happened to the Houston Astros' Hacker?*, https://www.si.com/mlb/2018/10/04/chris-correa-houston-astros-hacker-former-cardinals-scouting-director-exclusive-interview, October 4, 2018.

52. Schmidt, Michael. *Cardinals Investigated for Hacking into Astros Database*, https://www.nytimes.com/2015/06/17/sports/baseball/st-louis-cardinals-hack-astros-fbi.html?smid%253D=tw-nytsports&_r=0, June 16, 2015.

53. Passan, Jeff. *The Real Threat for Cardinals is How High Investigation Could Reach*, https://sports.yahoo.com/news/the-real-threat-for-cardinals-is-how-high-investigation-could-reach-191525747.html, June 16, 2015.

54. Nightengale, Bob. *Cardinals Flap Highlights New Unwritten Rule: Change Your Password*, https://www.usatoday.com/story/sports/mlb/2015/06/16/cardinals-astros-hack-fbi-investigation/28837817/, June 16, 2015.

55. Gartland, Dan. *Yankees Accuse Red Sox of Using Apple Watch to Steal Signs*, https://www.si.com/mlb/2017/09/05/yankees-red-sox-sign-stealing-apple-watch, September 5, 2017.

56. Schmidt, Michael. *Boston Red Sox Used Apple Watches To Steal Signs Against Yankees*, https://www.nytimes.com/2017/09/05/sports/baseball/boston-red-sox-stealing-signs-yankees.html?smid=tw-nytsports&smtyp=cur&_r=0, September 5, 2017.

57. Associated Press. *Hitters Knew Pitches in Stretch Drive*, http://www.espn.com/classic/s/2001/0201/1054936.html, November 19, 2003.

58. Verducci, Tom. *Why MLB Issued Historic Punishment to Astros for Sign Stealing*, https://www.si.com/mlb/2020/01/13/houston-astros-cheating-punishment, January 13, 2020.

59. Davis, Scott. *Astros Players are Already Getting Hit By Pitches Repeatedly in Spring Training*, https://www.businessinsider.com/astros-players-hit-by-pitches-spring-training-2020-2, February 27, 2020.

60. Baccellieri, Emma. *Sign Stealing Has Long Been a Part of MLB. It's Not Going Anywhere*, https://www.si.com/mlb/2019/11/13/sign-stealing-baseball-history, November 13, 2019.

61. Axisa, Mike. *We Now Know Extent of Cardinals Hack and the Unprecedented Penalties from MLB*, https://www.cbssports.com/mlb/news/we-now-know-extent-of-cardinals-hack-and-the-unprecedented-penalties-from-mlb/, January 30, 2017.

62. Verducci, Tom. *Baseball's Fight to Reclaim Its Soul*, https://www.si.com/mlb/2020/02/29/baseball-future-technology-astros, March 3, 2020.

63. Brown, Tim. *Everything's Under Such a Microscope Now: MLB Rules Crackdown Comes with Confusion*, https://sports.yahoo.com/mlb-rules-crackdown-pine-tar-sign-stealing-angels-astros-confusion-235800752.html, March 6, 2020.

64. Cwik, Chris. *MLB Will Crack Down on Pitchers Using Foreign Substances on the Ball in 2020*, https://sports.yahoo.com/report-mlb-will-crack-down-on-pitchers-using-foreign-substances-on-the-ball-in-2020-211419675.html, February 26, 2020.

65. Verducci, Tom. *Clean It Up. It Must Stop: MLB is in an Ethical Crisis*, https://www.si.com/mlb/2020/01/16/astros-cheating-scandal-baseball-crisis, January 17, 2020.

66. Stanford Encyclopedia of Philosophy. *Kant's Moral Philosophy*, https://plato.stanford.edu/entries/kant-moral/, July 7, 2016.

67. Pecorino, Philip. *Ethical Traditions: The Categorical Imperative*, https://www.qcc.cuny.edu/socialsciences/ppecorino/medical_ethics_text/Chapter_2_Ethical_Traditions/Categorical_Imperative.htm#:~:text=The%20Categorical%20Imperative%20is%20devised,%2D%2D%20that%20is%20our%20duty.&text=For%20Kant%20the%20basis%20for,the%20intention%20or%20the%20will.&text=Kant%20expressed%20this%20as%20the%20Categorical%20Imperative, 2002.

68. Mushnick, Phil. *New Rules Just Made NFL Games Even Slower*, New York Post, March 29, 2019.

69. Baseball Reference.com. *2008 MLB Team Statistics*, https://www.baseball-reference.com/leagues/MLB/2008.shtml, October 28, 2020.

70. Frost, Robert. *A Day of Prowess*, https://vault.si.com/vault/1994/07/11/a-day-of-prowess, July 11, 1994.

The last inning

Chapter Outline

Human beings have always had a penchant for numbers and counting. The most primitive forms of counting sought to match physical items or observations with a familiar set of objects. These sets of objects often included fingers, toes, rocks, and notches carved into sticks. There are vestiges of this rudimentary form of tracking quantities deeply rooted in the English language. The Oxford English Dictionary (OED) defines the word digit as "a person's finger (or thumb), or toe." Additionally, the OED defines the word digit as "a whole number less than ten; any of the nine or (including zero) ten Arabic numerals representing these, a series of which is used to represent other numbers in decimal notation."[1] Native Americans, most notably the Mayans, would often use their fingers and toes, or digits, for counting quantities, resulting in a vigesimal system based on groups of 20.[2] After 20 fingers and toes had been exhausted for counting, a notch was scored into a stick and the process was started over. While using the word score was originally used to denote a quantity of 20, as made famous by Abraham Lincoln's "Gettysburg Address," from April to October baseball teams across the country can be seen scoring runs to the delight, or dismay, of their fans.

Given human proclivity for numbers and counting, it is logical that we apply this interest as we seek answers that explain the outcomes of competitive games. Sports broadcasts from every sport contain a steady stream of statistics, as commentators attempt to harness our attention by catering to our numbers intrigue. What makes baseball different? First, baseball is discrete in nature and less fluid than many other sports. Each pitch results

Sabermetrics. DOI: http://dx.doi.org/10.1016/B978-0-12-822345-1.00009-X

in an outcome, and for over a century it has been relatively easy to capture a variety of basic statistics. Furthermore, many of the statistics can be directly attributed to individual players with minimal subjectivity. Fans of the game have many of these numbers deeply engrained in their psyche, by which all others will be measured against. When it comes to batters, numbers such as .406, 73 (or 61 for most skeptics), and 191 have the greatest meaning. Even though many regard batting average as an antiquated metric, mention the number .406 and most baseball fans will immediately know that you are referencing Ted Williams's storied 1941 season. The year 1941 was famous for more than Ted Williams's batting average. In the same year that Ted Williams hit .406, Joe DiMaggio had a 56-game hitting streak, both modern-era records that have outlived generations of baseball players. There is good reason for that. During Joe DiMaggio's 56-game hitting streak, he maintained a batting average of .408 with only five strikeouts. Even with this incredible batting average, and with a few simplifying assumptions, the probability of DiMaggio hitting in 56 consecutive games is approximately 0.0006, or 6/10,000.[3]

While the casual observer can appreciate anomalies such as Ted Williams's and Joe DiMaggio's 1941 season, statisticians seek to quantify phenomena in a game with seemingly innumerable variables. The game of baseball, and commensurately the field of baseball statistics, has evolved greatly since the mid-1800s. Henry Chadwick is credited for recording the first published box score in 1859 when the Brooklyn Stars defeated the heavy favorite Brooklyn Excelsiors.[4] The box score, printed in the *New York Clipper*, contained runs, hits, put-outs, assists, and errors. Although often criticized, since 1859 the box score has been simplified to include just runs, hits, and errors. That is not the case with most other baseball metrics. Rule changes and technological advances have changed how many view baseball statistics. What was formerly used for the purpose of reading pleasure has gradually been replaced with the purpose of teams, players, and fans gaining a competitive advantage. As a result, the friendly confines have become increasingly cutthroat.

The evolution of baseball statistics

Henry Chadwick, who many consider the father of sabermetrics, began his career as a cricket reporter for the New York Times. As a British immigrant, whose family settled in Brooklyn, New York in the late 1830s, Henry Chadwick was an amateur cricket player whose knowledge of the game and its rules superseded his reputation as a player. As "base ball" began to flourish in the 1840s and 1850s, Chadwick developed a system of abbreviations and codes that summarized games for teams in New York.

Henry Chadwick was not the only reporter to capture highlights of a baseball game for the purpose of publication. William Cauldwell of the *New York Times* began publishing rudimentary statistics for runs and outs in baseball games as early as 1845. A more formalized process began circulating as a result of the 1861 issue of *Beadle's Dime Base-Ball Player*, where Chadwick published what is widely acknowledged as the first standardized box score. Remnants of the Chadwick's 1861 box score still exist in box scores today. Most notably, the letter K, which Chadwick used to signify "struck out," is still the formal abbreviation for a strike out. The statistics that Chadwick developed were used for entertainment; their purpose was to tell a story and allow readers to understand how a game played out.[5] By 1864, Chadwick noted that as players became employees of businesses, statistics were the only fair way to describe players' abilities and were becoming increasingly important. Common statistics of hits per game or runs per game fell short of describing a player's talent or contribution. By 1870, Henry Chadwick was receiving a steady stream of recommendations for new metrics for measuring player performance, a practice that has continued for over 150 years.

The most notable recommendation came from a man named H.A. Dobson. As Henry Chadwick was developing a statistic to account for total bases (one for a single, two for a double, three for a triple, and four for a home run); a concept adopted into the modern-day slugging percentage metric, Dobson suggested that these measures of hits per game, runs per game, and total bases were skewed in favor of players who had more at-bats due to their position in the batting order. To combat this disparity, Dobson suggested that the number of hits a player has should be divided by the number of at-bats. In the 1872 *Beadle's Dime Base-Ball Player*, Chadwick gave his full support to this new metric, stating that, "according to man's chance, so should his record be."[6] This measure has stood the test of time as one of the most used statistics to describe batters, the batting average.[7]

One example of the evolution of sabermetrics can be seen in the statistics published on baseball cards. One of the most recognized baseball card companies, Topps, began publishing statistics on baseball cards in 1952. The 1952 Topps cards had two lines of statistics on each player, the previous season's statistics and career statistics. For batting statistics, Topps included games, at-bats, runs, hits, home runs, runs batted in (RBI), and batting average. Fielding statistics included put-outs, assists, errors, and fielding average. For pitchers, Topps included statistics for games played, innings pitched, winning percentage, hits allowed, runs allowed, strike outs, walks, earned runs, and ERA. The 407-card set of 1952 Topps baseball cards, sold for a nickel in packs of six that also included a stick of bubble gum, quickly became an obsession with children across the country.[8]

Over the next seven decades, baseball card statistics would see minor adjustments, such as including complete year-by-year statistics in 1957. By the mid-1980s, Topps had eliminated fielding statistics, but in addition to the original 1952 batting statistics, Topps added doubles, triples, stolen bases, strikeouts, and slugging percentage. This was perhaps an indication of baseball's impending performance-enhancing drug (PED)-fueled obsession with offensive statistics. By 2020, Topps baseball card statistics for batting added on-base plus slugging (OPS), yet another tacit acknowledgement that many baseball fans were interested in players who hit for extra bases, namely home runs. Arguably, pitching statistics have seen the greatest evolutions. This can be attributed to the fact that in the earliest baseball, pitchers were responsible for delivering an underhand toss as a means to let batters hit the ball. In addition to the pitching statistics listed on the 1952 Topps baseball cards, the 1985 cards included shutouts, earned run average (ERA), saves, games started, and complete games. By 2020, Topps had included the modern statistics of hits per inning pitched (WHIP) and wins above replacement (WAR). The WAR statistic did not become part of *Baseball-Reference's* statistical toolbox until 2009.

Major League Baseball (MLB) maintains a glossary to document modern sabermetrics. As technology and computational capacity has improved, so has the MLB list of advanced baseball statistics shown in Table 9.1.[9] Despite the growing number of advanced statistics used in baseball, there is no consensus on which are most important. Formal statistics have crept their way into traditionally subjective measurements. For example, the SABR Defensive Index (SDI), an index that combines multiple defensive metrics, is used to determine the list of qualified candidates for the Gold Glove Award, which is then voted on by managers and coaches. Well-known accolades such as the most valuable player (MVP) and Hall of Fame remain subjective assessments but grounded in many of the traditional metrics. How we evaluate baseball players, and even who we evaluate, has changed dramatically since the beginning days of the game.

Changing the rules

Rule changes throughout the late-1800s helped shape the modern game of baseball. Perhaps the biggest change was how the ball was delivered to the batters, which led to pitchers turning into throwers. In 1857, 16 New York area clubs organized the National Association of Base Ball Players (NABBP) to formalize rules for what became the *knickerbocker era* of baseball. The NABBP rules stated that pitchers must "deliver the ball as near as possible over the center of the home base," and that "the ball must be pitched, not jerked

LAWRENCE PETER BERRA
191
Catcher: New York Yankees   Home Woodcliff Lake, N. J.
Born: May 12, 1925, St. Louis, Mo. Eyes: Brown Hair: Black
Ht.: 5'8''   Wt.: 183   Bats: Left   Throws: Right

The "Most Valuable Player" in the American League in 1951. Yogi became a Yankee at the end of the '46 season. He spent 1943 at Norfolk, 1944-45 in the Navy and 1946 at Newark, where he hit .314. A power-hitter, he drove in 98 runs and hit .305 in '48 and clouted 20 Homers and had 91 RBI's in '49. In '50, he hit .322, and 28 Home Runs, 124 RBI's and made the Major League All-Star Team. Yogi is the first player to ever pinch-hit a Homer in a World Series (1947).

				MAJOR LEAGUE BATTING RECORD					**FIELDING RECORD**			
	Games	At Bat	Runs	Hits	Home Runs	RBI	Batting Average	Put-outs	Assists	Errors	Field. Avg.
PAST YEAR	141	547	92	161	27	88	.294	693	82	13	.984
LIFE-TIME	623	2343	381	701	102	459	.299	2739	270	51	.983

© T. C. G.   ☆ **TOPPS BASEBALL** ☆   PRTD. IN U.S.A.

EDMUND WALTER LOPAT
57
Pitcher: New York Yankees   Home: Hillsdale, N. J.
Born: June 21, 1918, New York, N.Y. Eyes: Hazel Hair: Red
Ht.: 5'10'' Wt.: 182 Throws: Left Bats: Left

One of the Big Leagues' best curve-ball pitchers, Ed became a 20-game winner in 1951 for the first time since he started in organized ball in 1937. Ed started out as a first baseman, but switched to the mound in his first year. He came up to the majors with the White Sox in 1944 and was traded to the Yanks in '48. As a Yankee, he's won 71 games in 4 seasons, losing only 38. In the winter, Ed runs a Baseball School in Florida.

					MAJOR LEAGUE PITCHING RECORD						
	Games	Pitched Innings	W	L	Pct.	Hits	Runs	Strike Outs	Bases on Balls	Ern. Runs	Avg. Run Ern.
PAST YEAR	31	235	21	9	.700	209	86	93	71	76	2.91
LIFE-TIME	243	1806	121	87	.582	1821	760	665	507	653	3.25

© T. C. G.   ☆ **TOPPS BASEBALL** ☆   PRTD. IN U.S.A.

1952 Topps Baseball Cards (Courtesy Topps).

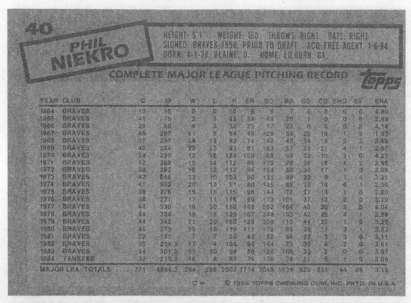

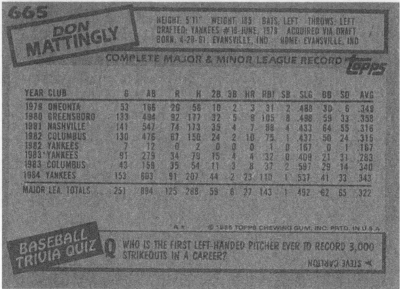

1985 Topps Baseball Cards (Courtesy Topps).

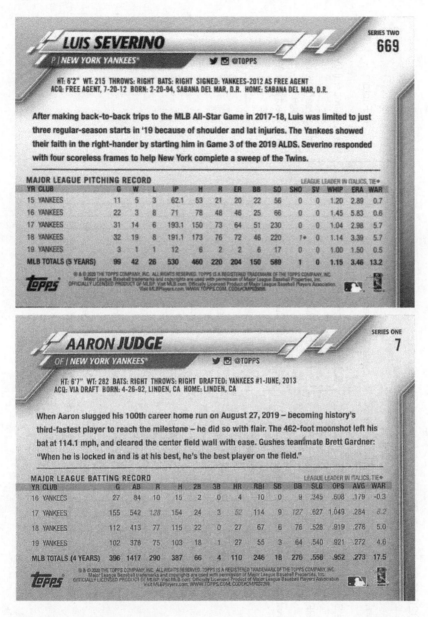

LUIS SEVERINO

P | NEW YORK YANKEES®

🐦 📷 @TOPPS

SERIES TWO
669

HT: 6'2" WT: 215 THROWS: RIGHT BATS: RIGHT SIGNED: YANKEES-2012 AS FREE AGENT
ACQ: FREE AGENT, 7-20-12 BORN: 2-20-94, SABANA DEL MAR, D.R. HOME: SABANA DEL MAR, D.R.

After making back-to-back trips to the MLB All-Star Game in 2017-18, Luis was limited to just three regular-season starts in '19 because of shoulder and lat injuries. The Yankees showed their faith in the right-hander by starting him in Game 3 of the 2019 ALDS. Severino responded with four scoreless frames to help New York complete a sweep of the Twins.

MAJOR LEAGUE PITCHING RECORD

LEAGUE LEADER IN ITALICS, TIE♦

YR CLUB	G	W	L	IP	H	R	ER	BB	SO	SHO	SV	WHIP	ERA	WAR
15 YANKEES	11	5	3	62.1	53	21	20	22	56	0	0	1.20	2.89	0.7
16 YANKEES	22	3	8	71	78	48	46	25	66	0	0	1.45	5.83	0.6
17 YANKEES	31	14	6	193.1	150	73	64	51	230	0	0	1.04	2.98	5.7
18 YANKEES	32	19	8	191.1	173	76	72	46	220	1♦	0	1.14	3.39	5.7
19 YANKEES	3	1	1	12	6	2	2	6	17	0	0	1.00	1.50	0.5
MLB TOTALS (5 YEARS)	99	42	26	530	460	220	204	150	589	1	0	1.15	3.46	13.2

AARON JUDGE

OF | NEW YORK YANKEES®

🐦 📷 @TOPPS

SERIES ONE
7

HT: 6'7" WT: 282 BATS: RIGHT THROWS: RIGHT DRAFTED: YANKEES #1-JUNE, 2013
ACQ: VIA DRAFT BORN: 4-26-92, LINDEN, CA HOME: LINDEN, CA

When Aaron slugged his 100th career home run on August 27, 2019 – becoming history's third-fastest player to reach the milestone – he did so with flair. The 462-foot moonshot left his bat at 114.1 mph, and cleared the center field wall with ease. Gushes teammate Brett Gardner: "When he is locked in and is at his best, he's the best player on the field."

MAJOR LEAGUE BATTING RECORD

LEAGUE LEADER IN ITALICS, TIE♦

YR CLUB	G	AB	R	H	2B	3B	HR	RBI	SB	BB	SLG	OPS	AVG	WAR
16 YANKEES	27	84	10	15	2	0	4	10	0	9	.345	.608	.179	-0.3
17 YANKEES	155	542	128	154	24	3	52	114	9	127	.627	1.049	.284	8.2
18 YANKEES	112	413	77	115	22	0	27	67	6	76	.528	.919	.278	5.0
19 YANKEES	102	378	75	103	18	1	27	55	3	64	.540	.921	.272	4.6
MLB TOTALS (4 YEARS)	396	1417	290	387	66	4	110	246	18	276	.558	.952	.273	17.5

2020 Topps Baseball Cards (Courtesy Topps).

Table 9.1 2020 Major League Baseball Advanced Statistics

Defense	Offense	Pitching	Team
Defensive Efficiency Ratio (DER)	Batting Average on Balls in Play (BABIP)	Adjusted Earned Run Average (ERA+)	Ballpark Factor
Defensive Runs Saved (DRS)	Isolated Power (ISO)	Baserunners per Nine Innings Pitched (MB/9)	Magic Number (MN)
Range Factor (RF)	Late-inning Pressure Situation (LIPS)	Bequeathed Runners (BQR)	Pythagorean Winning Percentage
Ultimate Zone Rating (UZR)	On-base Plus Slugging Plus (OPS+)	Bequeathed Runners Scored (BQR-S)	Win Expectancy (WE)
	Pitches per Plate Appearance (P/PA)	Expected Fielding Independent Pitching (xFIP)	
	Plate Appearances per Strikeout (PA/SO)	Fielding Independent Pitching (FIP)	
	Runs Created (RC)	Fly-ball Rate (FB%)	
	Weighted Runs Above Average (wRAA)	Game Score	
	Weighted On-base Average (wOBA)	Ground-ball Rate (GB%)	
	Weighted Runs Created Plus (wRC+)	Hits per Nine Innings (H/9)	
	Win Probability Added (WPA)	Home Runs per Nine Innings (HR/9)	
	Wins Above Replacement (WAR)	Home Run to Fly Ball Rate (HR/FB)	
		Inherited Runs Allowed (IR-A)	

Inherited Runs Allowed Percentage (IR-A%)

Innings per Start (I/GS)

Leverage Index (LI)

Line-drive Rate (LD%)

Pitches per Inning Pitched (P/IP)

Pitches per Start (P/GS)

Pop-up Rate (PO%)

Run Support per Nine Innings (RS/9)

Runs Allowed per Nine Innings Pitched (RA9)

Skill-interactive Earned Run Average (SIERA)

Strikeout Rate (K%)

Strikeouts per Nine Innings (K/9)

Strikeout-to-walk Ratio (K/BB)

True Earned Run Average (tERA)

Walks per Nine Innings (BB/9)

Walk Rate (BB%)

or thrown to the bat." This is where the term *pitcher* originated. For this reason, pitching statistics were notably absent until almost 30 years after the formation of the NABBP. In 1871, abandoning their official amateur status, the NABBP became the National Association of Professional Base Ball Players (NAPBBP), known as the *National Association*. In 1876, several clubs from the National Association created the National League of Professional Baseball Clubs, known as the *National League*, which stands today as one half of Major League Baseball. The National League allowed overhand pitching in 1884, which led to the development and recording of many pitching statistics.

Overhand pitching was not the only notable rule change in the late 1800s; the 1880s and 1890s were filled with rule changes, many of which are still present in professional baseball. The distance from the mound to home plate, formerly home base, changed three times in the late 1880s. Originally, the distance from the pitcher to home base was 45 feet. In 1881, this distance was increased to 50 feet, and added that the distance was to be measured from the pitcher's front foot to home base. Pitchers during this time would often get a running start prior to delivering a pitch. When overhand pitching was allowed in 1884, the running start from a distance of 50 feet amplified the advantage pitchers got from throwing the ball overhand, even with batters allowed four strikes. For the next decade, the distance from home plate would move to 55 feet 6 inches, and the number of strikes allowed decreased to three. By 1893, the National League had settled on a pitching distance of 60 feet 6 inches, and batters were allowed four balls and three strikes per at-bat. This is still the distance from the pitching mound to home plate and an early indication that many rules in professional baseball are reactionary as an attempt to restrict a player's ability to gain a competitive advantage. There were several other notable changes to the rules during this time period. By 1900, most players in the field wore gloves and fielding the ball off one bounce was no longer considered an out. The year 1900 begins what most baseball historians consider the *modern era*, due to rule changes generally codified and resembling the rules of current professional baseball and due to the formation of the American League at the turn of the century.

As the increased distance between pitchers and batters gave pitchers less of an advantage from throwing the ball fast, they resorted to developing different types of pitches that took advantage of the full 60 feet 6 inches. There is much debate on who invented the curveball. The Baseball Hall of Fame plaque for Candy Cummings states that he threw the first curveball in 1867, but exotic pitches had minimal benefit before overhand pitching was made legal. By the early-1900s curveballs, spitballs, sinkers, sliders, screwballs, and knuckleballs were commonplace in professional baseball. Additionally, in 1900 home plate was widened from 12 inches to 17 inches and changed to a pentagonal shape. The rules of professional baseball were considered "pitcher friendly" and games were lower scoring than ever before. This was apparent in statistics from this time period. In 1901, the St. Louis Cardinals led the league in home runs; they hit just 39. As such, many refer to the period of time from 1900 to 1919 as the *dead ball era*. One rule change and a handful of players turned the *dead ball era* into the *live ball era* in 1920.

The year 1920 was tragic in the world of professional baseball. On August 16, 1920, in a game between the New York Yankees and Cleveland Indians, a ball thrown by Yankees pitcher Carl Mays struck Indians shortstop Ray Chapman in the head. Chapman died the next day. Ray Chapman was not the first baseball player to be killed by a pitch to the head. The first documented case was Dayton Veterans second baseman Charles Pinkney, who was struck by a pitch and killed on September 14, 1909. However, it was the death of Ray Chapman that led to a series of changes. Many attribute Chapman's death to the widespread tampering of baseballs.[10] The ball was brown from overuse, likely made worse by pitchers' frequent use of spitballs, and many believe that Chapman never saw the ball coming toward his head. While Chapman's death is what drove a helmet mandate in the late-1950s, more immediate changes impacted the actions of pitchers and umpires.[11] Pitchers were banned from tampering with balls and umpires were required to replace overused balls with white baseballs to make them more visible to batters. The rule changes of 1920 had an immediate impact. Babe Ruth hit 11 HRs in 1918, 29 HRs in 1919, 54 HRs in 1920, and 59 HRs in 1921.

The period of time between 1920 and 1969 lacked many significant changes to rules. Most new rules during this time regulated the size of ball parks. In 1925, the minimum distance from home plate to the fence was set at 250 feet. In 1959, Major League Baseball required that all new stadiums measure 325 feet from home plate to the left and right foul poles, and 400 feet from home plate to straightaway center. There are still fields that do not meet the standards enacted in 1959, the most iconic of which is the "green monster" at Fenway Park. Built in 1912, the leftfield wall of Fenway Park is just 310 feet from home plate, but measures just over 37 feet in height. The idiosyncrasies of baseball parks have an effect on statistics. To compensate for this variability, many modern statistics incorporate a park factor (PF), which is:

$$PF = \frac{\frac{HRS + HRA}{HG}}{\frac{RRS + RRA}{AG}},$$

where HRS is runs scored at home, HRA is runs allowed at home, HG is home games played, RRS is runs scored on the road, RRA is runs allowed on the road, and AG is away games played. The dimensions of Fenway Park and the elevation of Coors Field regularly place these two ballparks in the top ten baseball fields in terms of park factor.

Seemingly minor adjustments to the playing field can have major effects on statistics. In 1968, MLB saw a year in which 49 pitchers finished the season with a sub-3.00 ERA (excluding pitchers with less than 1 inning pitched per game), and the league-wide ERA average was 2.98. The 1968 MLB All-Star Game was arguably the most boring All-Star Game in history. The NL starting pitcher, Dodgers' Don Drysdale, pitched 58 consecutive innings without a run, to include 6 consecutive shutouts. One American League batter entered the game with a batting average over .300, Carl Yastrzemski, with a .301 batting average. While the game at the Houston Astrodome made history for being the first All-Star Game played indoors, from an offensive standpoint, the 1968 All-Star Game was predictably lackluster. The National League won the game 1-0 on a Willie Mays run in the first inning. Don Drysdale was the winning pitcher. The 1968 season ended with Bob Gibson having a 1.12 ERA, the lowest in MLB since 1914, and continued to be the lowest in over 100 years. Something had to change. In 1969, MLB lowered the mound height from 15 to 10 inches, and it worked. In 1969, the league-wide ERA was 3.61, and just 24 pitchers had a sub-3.00 ERA. Seven players had 40 or more HRs in 1969, compared to just one in 1968. Eighteen players finished the 1969 season with a batting average over .300, compared to just six in 1968. While many attribute the offensive production of 1969 to the change in mound height, another subtle change may have been a contributing factor.

As of 2020, MLB defines the strike zone as "the area over home plate from the midpoint between a batter's shoulders and the top of the uniform pants—when the batter is in his stance and prepared to swing at a pitched ball—and a point just below the kneecap," but this definition has changed numerous times throughout the modern era.[12] From 1963 to 1968, the strike zone was defined as the area over the plate between the batter's shoulders and his knees. In 1969, the strike zone changed from the batter's armpits to the top of the knees, making the strike zone slightly smaller. The smaller strike zone of 1969 coincided with the decreased mound height, both of which resulted in a higher offensive production. The smaller strike zone introduced in 1969 remained in place through the 1987 season. In 1988, MLB introduced an even smaller zone that went from the midpoint between the shoulders and the top of the uniform pants (in the vicinity of the lettering across the front of the uniform), to the top of the knees. In 1995, the strike zone was expanded from the top of the knees to just below the kneecaps, but this appeared to have very little effect on offensive statistics. In 1994, there was a league batting average of .270 and an OPS of .763, whereas 1995 saw a league batting average of .267 and an OPS of .755. Rule changes add an extra layer of complexity when comparing player statistics across multiple years, in particular when playing careers span numerous, notable rule changes.

The elusive competitive advantage

The earliest baseball statistics were primarily published for fans, but this shifted to evaluating players, managerial decisions, and an opportunity to gain the elusive competitive advantage. While baseball is a game, it is also a business. In 2020, *Forbes* estimated the value of the New York Yankees at $5,000,000,000. Given the popularity of Henry Chadwick's 1859 *New York Clipper* box score, the business aspect of professional baseball has pushed the frontiers of player evaluation and forecasting. This behind-the-scenes number crunching gained the intrigue of many casual fans in 2003 with the release of Michael Lewis's *Moneyball: The Art of Winning an Unfair Game.*

In *Moneyball,* Michael Lewis tells the story of the Oakland Athletics' success in professional baseball, despite having a payroll that was only about one-third that of the largest clubs. In 2001, Oakland had the second lowest payroll in MLB, but they finished the season with 102 wins and made it to the American League Division Series. That year there were 21 teams with a higher payroll than the Athletics that didn't even make the playoffs. Their secret? According to Lewis, "That answer begins with an obvious point: in professional baseball it still matters less how much money you have than how well you spend it." General Manager Billy Beane embraced unconventional analytics in order to gain a competitive advantage over other ball clubs. One example of this was when Beane preferred on-base percentage and slugging percentage over batting average as a better measures of a player's offensive contribution. Based on what he believed was an overevaluation of his own talent coming out of high school, Beane also preferred to draft players out of college who had a greater body of data collected in them. These players were viewed as less risky and gauging their true talent level was more predictable. Beane believed that statistics allowed the Athletics to reduce traditional biases, or what Lewis referred to as "sight-based scouting prejudices."[13] For pitching statistics, Beane preferred statistics such as high ground ball to fly ball ratios and the newly-developed defense-independent pitching statistic (DIPS) over more traditional measures such as arm strength. The approach gave the Oakland Athletics a clear advantage as they made the playoffs for four consecutive years (2000 to 2004) with one of the lowest MLB payrolls. Sabermetrics are a valuable resource when building teams, but the search for a competitive advantage does not end once a team is assembled.

On June 14, 1870, Bob Ferguson of the Brooklyn Atlantics became the first switch hitter in professional baseball. In 1870, pitchers were still tossing, or *pitching*, the ball underhand, so Ferguson's reason for switch hitting was not for the same reasons hitters switch hit today. Ferguson was a right-handed pull hitter and he decided to hit from the left side of the plate to avoid hitting the ball to the Cincinnati Red Stockings' shortstop George Wright. Down one run in the bottom of the 11th inning with a runner on base, Ferguson not only drove in the runner on base, but scored the winning run on a throwing error.[14] Whether Ferguson's hit that ended the Red Stockings' 81-game winning streak can be directly attributed to switch hitting will never be known, but there was enough evidence of its effectiveness that managers began adjusting their defenses accordingly.

Origins of the infield shift can be traced as early as the 1920s, but the practice did not become mainstream until the 1940s. It was on July 14, 1946 in a game between the Cleveland Indians and the Boston Red Sox, when Cleveland player-manager Lou Boudreau decided to move both the shortstop (himself) and the third baseman to the right side of second base when facing one of the most prolific hitters of all time, Ted Williams.[15] Ted Williams was a pull hitter who found much of his success hitting balls down the right field line. The first time Ted Williams faced the Indians' infield shift, he grounded out to Boudreau. From that moment on, Williams, who refused to adjust his swing, would regularly face teams that shifted their defense towards the right field. However, the infield shift did not become a common practice until the 2000s.

From 2010 to 2019, use of the infield shift in MLB increased nearly by 2000%. In 2019 alone, teams employed a defensive shift over 46,000 times, compared to just 2350 in 2010.[16] The effectiveness of non-traditional defensive alignments is still in question. How you define effectiveness, or even how you define a shift, contributes to that answer. Mark Simon of Sports Information Solutions presented his findings at the 2018 SABR Analytics Conference. First, he set out to define what constitutes and infield shift, breaking the concept into two categories, a partial and a full shift. A partial shift was defined as one player (typically the shortstop) aligning in the middle of the field, whereas a full shift was when a team had three infielders on one side of second base. Since the shift has very little effect on fly balls and home runs, Simon looked specifically at defensive performance against ground balls and line drives. His conclusion was, the infield shift works. While there are differences in effectiveness from player to player, in 2017, the league-wide batting average against the full shift was .241, against the partial shift it was a .269, and the batting average against no shift was .271.[17] However, Russell Carlton of Baseball Prospectus came to a different conclusion. He estimated that from 2015 to 2017, a full infield shift decreased the number of singles by 493, but increased the number of walks by 574. One of his conclusions was, if the purpose of the shift was to prevent players from reaching first base, the full shift does not work. Regardless of the differing opinions, the infield shift remains a common practice in the modern game of baseball and its efficacy will be debated for years to come.[18]

Perhaps baseball statistics have become an example of Occam's razor. The data driving the analysis by Russell Carlton and Mark Simon was made available by the cutting-edge technology of Stat Cast. Is the abundance of data only complicating analysis that can be done using rudimentary statistics? For example, the players with eight of the top ten highest career batting averages did so from the left side of the plate. Four of these players were biologically left-handed (Lefty O'Doul, Tris Speaker, Dan Brouthers, and Babe Ruth). Assuming 10% of the population was left-handed, a figure generally agreed upon in the scientific community, what is the probability that a random sample of 10 people would yield at least 4 left-handed individuals?[19] That probability is an astonishingly low 1.287%. It would be easy to surmise that left-handed batters are more productive. However, it wasn't until September 11, 2020, in a game at Tropicana Field against the Boston Red Sox, that Kevin Cash, manager of the Tampa Bay Rays, became the first manager to field a starting lineup with nine left-handed batters. Not all of the batters were biologically left-handed, but all nine hit exclusively from the left side of the plate. The Rays won the game 11-1.

Always an asterisk

As a result of the PED-fueled home run races in the late 1990s and early 2000s, many of the prominent voices in baseball placed an asterisk next to records where they believed players gained an unfair advantage. This is not a dilemma unique to the modern game of baseball. There is an asterisk associated with nearly every generation of professional players and any statistics or records must be taken in the context of the period in which they were recorded.

As Babe Ruth littered headlines in the 1920s and attracted thousands to ballparks across the country, there were names such as Josh Gibson, Satchel Paige, and Mule Suttles who deserved their share of headlines, but were denied for one reason: the color of their skin. Racial integration did not occur in MLB until April 15, 1947, the day that Jackie Robinson famously made his appearance at Ebbets Field for the Brooklyn Dodgers. That struggle for racial integration began long before 1947. Black baseball teams date back to the late 1800s, but the seven leagues referred to as the *Negro Major Leagues* from 1920 to 1950 are typically recognized as organizing the highest quality programs and have rightfully taken their place in the record books. On December 16, 2020, MLB Commissioner Robert Manfred Jr. announced that the Negro Leagues that began 100 years prior would officially be recognized as Major League and their statistics would be included as part of MLB. How those records are viewed must be taken in context of the timeframe. For example, a player with

32 HRs in a season would be considered above average. However, in 1926, Mule Suttles hit 32 HRs in a full 89-game season for the St. Louis Stars of the Negro National League. Even when compared against MLB players in their 154-game season of 1926, 32 HRs places Mule Suttles second, only behind Babe Ruth who had 47 HRs in 1926. In that year, Mule Suttles hit 0.36 HRs per game, compared to Babe Ruth who hit 0.31 HRs per game. How should Mule Suttles's performance in 1926 be compared to other league leaders? The answer to this nontrivial question has applicability as we analyze other statistics with asterisks.

Less than two decades into the modern era of baseball, the United States sent nearly 3,000,000 service members overseas to serve in the "war to end all wars," World War I. The war had a significant impact on professional baseball. The Society of American Baseball Research identified 777 players, umpires, scouts, and league officials who served during World War I.[20] The war interrupted the careers of several notable players and managers. Future Hall of Famers Christy Mathewson and Ty Cobb served in World War I, along with manager Branch Rickey, later famous for signing Jackie Robinson. The impacts were felt across MLB. Of those who served, 196 players played in a game during the war. The 1918 season was reduced by approximately 30 games as a result of the US government's "work or fight" mandate, which required American men to either take a job that directly supported the war effort or enlist in the armed forces. That year Ty Cobb won the American League batting title, hitting .382 with less than 200 hits.

The war to end all wars didn't, and a little more than two decades after World War I, the United States declared war against Japan after the bombing of Pearl Harbor in 1941. America's involvement in World War II had begun. Prior to the bombing of Pearl Harbor, 1941 was one of the most exciting years in professional baseball. Joe DiMaggio had a 56-game hitting streak and Ted Williams batted .406. Even after the United States declared war against Japan, MLB continued to play 154-game seasons throughout World War II. Many players interrupted their playing careers to answer the nation's call. Joe DiMaggio would enlist in the Army Air Corps in 1943, Yogi Berra would serve as a gunner's mate on a landing craft during the D-Day invasion, and Ted Williams would serve as a pilot during both World War II and the Korean War. In all, more than 500 MLB players served in the armed forces during World War II, many of whom sacrificed years in their baseball prime to make selfless contributions to the war effort. Two MLB players made the ultimate sacrifice. Elmer Gedeon and Harry O'Neill were killed in action during World War II.[21]

Periods of war and conflicts after World War II saw fewer and fewer athletes. Just over 100 players served during the Korean War and the SABR identified 120 players, umpires, scouts, and league officials who served in Vietnam. The Hall of Fame bears the names of 71 war veterans that include one civil war veteran, 28 World War I veterans, 36 World War II veterans, and six Korean War veterans. There are no Vietnam veterans in the Hall of Fame. Wartime service undoubtedly affected career statistics. In the prime of his career, Ted Williams did not play in a single game for three years (1943 to 1945), only played 6 games in 1952, and played just 37 games in 1953, seemingly insurmountable asterisks for even the most storied playing career. Even given these interruptions, for his career, Ted Williams ranks 20th in runs scored, 20th in home runs, 15th in RBIs, 4th in walks, and was an All-Star for 17 seasons. While the 1940s and 1950s saw players selflessly set aside their own professional aspirations in defense of their country, half a century later MLB was dealing with players who would go to extreme measures to have their names recorded in the annals of baseball. The PED era had begun.

The use of PEDs in professional baseball was sadly predictable. In 1880, Henry Chadwick noted, "the present method of scoring the game and preparing scores for publication is faulty to the extreme, and it is calculated to drive players into playing for their records rather than for their side."[22] At the beginning of the 1998 season, Roger Maris's single season home run record of 61 had stood since 1961. That season, Sammy Sosa and Mark McGwire would capture the baseball world's attention as they hit baseballs over the fences at a rate never seen before. The 1998 season ended with McGwire shattering Maris's record with 70 home runs, followed closely by Sosa with 66. The 1999 season ended in a similar fashion with McGwire hitting 65 home runs and Sosa hitting 63. Skepticism finally hit an all-time high in 2001 when Barry Bonds hit 73 home runs. Steroids were on MLB's banned substance but were not tested until 2003, after players and owners agreed to a collective bargaining agreement the year prior.[23]

Subjecting players to mandatory testing for PEDs confirmed or dispelled many rumors about PEDs, but the tests did not predate 2003. Furthermore, even positive tests often created ethical dilemmas for fans, coaches, and sportswriters. It is something that Hall of Fame voters have been struggling with for over a decade. Not all instances of PED use are equivalent. Despite overwhelming circumstantial evidence, some players adamantly deny using PEDs. Other players admit using PEDs, but only on a limited basis. Some players admit to unknowingly using PEDs. Finally, and less controversially, there are those who admit to long-term PED use. Some notable players under a veil of suspicion who deny having used PEDs include Sammy Sosa and Roger Clemens. Sammy Sosa finished his career with 609 home runs and is the only player in baseball history to have three seasons with more than 60 home runs. Roger Clemens played 24 seasons, has 4,672 strike outs, was an 11 time All-Star, earned seven Cy Young awards, and has two WS championships. Even with those accolades, as of 2020 Roger Clemens has not received more than 61% of the Hall of Fame vote, well short of the required 75%.

Many players admit to unknowingly using steroids, the most famous of which is Barry Bonds. Bonds was a 14-time All-Star, a 7-time MVP, the single season home run record holder (73), and career home run record holder (762). Bonds admits to using steroids in 2004, but not knowing what they were at the time. Like Clemens, Bonds has not received more than 61% of the Hall of Fame vote. Gary Sheffield and Rafael Palmeiro, both with over 500 career home runs, also admit to using steroids, but not knowing what they were at the time. Palmeiro famously testified before Congress that he hadn't used PEDs, only to test positive just months later. Other players leave less doubt as to their use of PEDs. Manny Ramirez, a 12-time All-Star with 555 career home runs, has twice tested positive for PEDs. Alex Rodriguez, a 14-time All-Star and 3-time MVP who finished his career with 696 home runs, admitted three years of PED use. Mark McGwire, a 12-time All-Star with 583 career home runs, admitted 10 years of PED use. Many fans are quick to dismiss the statistics of players who knowingly used PEDs to gain a competitive advantage, but each situation is unique.

Andy Pettitte had a storied 18-year career, all with the New York Yankees. He holds the record for most playoff wins and playoff innings pitched; he has five World Series championships out of the eight World Series he appeared in. Pettitte was one of the names mentioned in the Mitchell Report in 2007 as having tested positive for PEDs.[24] Pettitte admitted using human growth hormone two to four times in 2002 to help heal an elbow injury. This posed an ethical dilemma for many fans. Is the use of PEDs okay if it is to help heal an injury? Are PEDs used for rehabilitation purposes any different than anatomical changes such as an ulnar collateral ligament reconstruction, commonly referred to as "Tommy John surgery?" It's an ethical debate, but baseball ethics are likely to become more broadly discussed in the midst of medical and technological advances.

Communication has become the latest asterisk in professional baseball. In high school dugouts across the country, coaches tell their players to watch the opposing team's third base coach and try to pick up on the signs given to batters and base runners. The opposing team often reciprocate by having base runners on second base try and pick up on the catcher's signal to the pitcher. Sign stealing has been a part of baseball for decades. However, in 2017 and 2018, the Houston Astros developed an elaborate system of cameras in center field with a feed directly into the Astros' dugout. Players in the dugout would then use audible cues to let the batters know what kind of pitch was coming next. The impact is difficult to quantify, but the Astros beat the Los Angeles Dodgers in the 2017 World Series and lost in the American League Championship Series in 2018. The Astros were hit with a $5,000,000 fine, and MLB took away their first and second round draft picks in the 2020 and 2021 drafts. MLB also suspended the Astros' coach and general manager for the 2020 season. Most fans would agree that the Astros crossed an ethical line with their over-the-top setup with cameras, monitors, and signaling schemes, but how far can you take that logic? If a team recognizes a "take" sign from the third base coach and tells the pitcher to throw a fastball over the middle of the plate, is that cheating? The larger, ethical question is what communication should be private? Catcher to pitcher? Coach to catcher? Coach to batter? If certain communications should be private, why doesn't MLB have microphones for coaches and speakers in the hats of pitchers, batters, and catchers? The National Football League (NFL) allows electronic communication between coaches and select players. An added bonus of allowing electronic communication is that it would speed up the game (a goal that comes up quite frequently as of late) since the pitcher does not have to sit through a catcher's signs before every single pitch. Pitchers currently have 12 seconds to deliver each pitch when no runners are on base.[25] Additionally, batters would not have to wait on signals from the third base coach before every pitch. Technological advances, and the opportunities that come with them, drive rule changes that are often met with resistance for those who insist on maintaining the "purity" of the game, but ethics are playing an increasingly important role in how baseball is played in the 21st century.

Technology, new rules, and the "footballization" of baseball

Video review was first introduced to MLB on a limited basis by commissioner Bud Selig on August 28, 2008 but has played an increasingly important role in how professional baseball games are officiated and how players are evaluated. In 2008, instant replay was limited to reviewing home run calls where there was a dispute over whether a ball was fair or foul, went over the fence, or hit an object. In 2008, and the five seasons that would follow, instant replay would be used 392 times, and 132 of those times the initial ruling was overturned.[26] In 2014, MLB announced a partnership with Hawk-Eye Innovations to expand the use of instant replay that included ground-rule doubles, fan interference calls, force outs, tag outs, fair/foul balls, dropped or trapped fly balls, whether or not a player scored before the third out, and scorekeeping issues.[27] Hawk-Eye's Statcast platform consisted of cameras and radars that were installed in every MLB stadium. It had an immediate impact. In 2014, there were 1280 challenged calls and 607 of the rulings were overturned. This trend continued for the next 5 years as shown in Table 9.2. The use of Statcast doesn't just enable fairness; it has the ability to improve the accuracy of the record books.

The expanded use of video review came too late for Detroit Tigers' pitcher Armando Galarraga. On June 2, 2010, Galarraga had a perfect

Table 9.2 Instant Replay Usage in MLB[28]

Year	Total Challenges	Rulings Overturned
2015	1,391	679
2016	1,605	815
2017	1,524	736
2018	1,502	702
2019	1,464	650

game taken away by the call of umpire Jim Joyce in what Galarraga later referred to as the "28-out perfect game." The Detroit Tigers were facing the Cleveland Indians when Indians rookie shortstop Jason Donald hit a ground ball to the Tigers' first baseman Miguel Cabrera. Cabrera came off the bag and tossed the ball to Galarraga, who was covering first base. Galarraga began celebrating, but Joyce called Donald safe. Replay clearly showed Donald was out. In a tearful apology after the game, Joyce admitted that he made the incorrect call. In a selfless, professional response, Galarraga accepted the apology and said that he would show his kids the tape of the game and explain that he threw a 28-out perfect game. Galarraga's perfect game would have been just four days after Max Halladay threw a perfect game, the shortest gap in the modern era, and 24 days after Dallas Braden's perfect game on May 9, 2010. From electronic strike zones, to pitch clocks, to expanded use of instant replay, the use of technology continues to be a source of debate between baseball traditionalists and those willing to sacrifice tradition in order to get correct calls.

In addition to the value added from an officiating standpoint, Statcast has opened new frontiers in the realm of player evaluation. In 2020, Statcast underwent an upgrade that incorporates optical sensors and 12 cameras, five of which are dedicated solely to the pitcher, in every MLB stadium. This technological advancement and ability for data collection enabled the advent of new statistics for player assessment. Statcast collects data on arm strength of players at all positions, runner speed between bases, the path a fielder takes to the ball, exit velocity and launch angle of batted balls, the time it takes a catcher to get the ball out of his glove, and the baseball revolutions per minute for pitches. New metrics based on these data include barrels (a combination of launch angle and exit velocity), catch probability, and expected batting average (probability that a batted ball is a hit). Many traditional metrics are designed to gauge a player's ability, while many of the new metrics allow coaches and analysts to determine whether a player is over or under performing.

Technological advances in the game of baseball have some critics questioning if the coming years will have any semblance to the game invented in the 1800s, but there are still rules and measurements that have stood the test of time. One example of this is the rule that dictates the size and weight of baseballs. In a game of inches with a myriad of specific metrics, there is a surprisingly large tolerance for the size and weight of official baseballs. Baseballs must be 9 to 9.25 inches in circumference and have a weight of 5 to 5.25 ounces. Additionally, baseballs are made from two pieces of horsehide or cowhide hand-sewn together with 108 double stitches. This has been true since 1876. Other rules change over time, and the year 2020 was a testbed for some of the more interesting rule changes of the modern era.

The COVID-shortened, 60-game season of 2020 brought with it changes that, in addition to the short season, substantially changed how the game was played. The designated hitter has been used in the American League since 1973. The National League has used a designated hitter when playing in American League parks for interleague play since 1997, but 2020 was the first year where a designated hitter was used universally. Many fans, who were already eager for live sports, expected an increase in offensive production for 2020, especially since National League pitchers batted just .129 in 2019, about half that of the rest of the league. However, even with a designated hitter, the National League hit .246 in 2020, compared to .251 in 2019. Additionally, the league OPS went from .753 to .746, and the runs per game dropped from 4.78 to 4.71. Many factors likely played into this decrease in offensive production, the most obvious being shortened training camps, but needless to say, the National League designated hitter experiment of 2020 was rather uneventful.

Pace of play took to the forefront of rule changes in 2020. In an effort to speed the pace of play, MLB introduced a rule that relief pitchers must face a minimum of three batters or complete a half-inning, but perhaps the most significant rule change in 2020 affected extra-inning games. Starting in the 10th inning, each team began with a runner on second base. The runner was the player who recorded the last out in the previous inning. The runner starting on second base was treated as reaching on an error, without actually recording an error. Therefore, in the event that the runner scored, the pitcher would not be charged with an earned run. The rule change was not without critics, with some claiming it was another example of what sabermetrics author Gabe Costa referred to as the "footballization" of baseball. In a 2013 article "By the Numbers: The World Series and the 'Footballization' of Baseball," Costa described how baseball games being played late at night in the fall, interleague play affecting the intrigue of the All-Star Game, and the length of the playoffs were reasons that some fans were turning to football to scratch their sports itch.[29] However, these rule changes impact player statistics. Managers did not have to go as deep in their bullpens and pitchers stayed fresh as the rule change resulted in just four games going beyond 11 innings in 2020, compared to 59 games of 12+ innings in 2019. Offensively, the runner starting extra innings changed how teams played. Prior to 2020, extra-inning contests often ended in the first team to hit a home run. A runner on second to start extra innings forced managers to shift their strategy to small ball, the strategy of advancing runners and attempting to score with singles or sacrifice fly balls. The rule change also created an unintended consequence: a pitcher could throw a perfect game and lose the game.

Strong pitching performances do not always translate to the record books. On June 28, 2008, Los Angeles Angels pitchers, Jered Weaver and Jose Arredondo, combined for a complete game against the Los Angeles Dodgers without giving up a hit. Weaver, who threw six no-hit innings, was the losing pitcher. The reason? The Dodgers won the game 1-0 after scoring an unearned run in the 5th inning. Furthermore, Weaver and Arredondo were not credited with a combined no-hitter because the MLB definition of a no-hitter stipulates that the pitchers must throw at least nine innings. The Angels were the visiting team, so Weaver and Arredondo did not pitch in the 9th inning. The rule changes for 2020 have the potential to exacerbate these inequities.

The MLB official definition of a perfect game is "when a pitcher (or pitchers) retires each batter on the opposing team during the entire course of a game, which consists of at least nine innings. In a perfect game, no batter reaches any base during the course of the game."[30] Using an extreme hypothetical, in 2020, a pitcher could go into the 10th inning having struck out 27 straight batters. In the 10th inning, a base runner would start the inning on second base. This still meets the definition of a perfect game because the runner on second base was not a batter who reached base, as stipulated in the definition of a perfect game. Highly improbable, but completely within the rules, that runner could reach third base and then score on a stolen base, sacrifice fly, sacrifice bunt, or fielder's choice. If the pitcher strikes out three batters in the 10th inning, he would have thrown a perfect game, having struck out 30 batters, and lose the game 1-0. He would quickly become one of the most famous pitchers to ever play the game. The rule changes of 2020 are not the first time MLB has sought out ways to increase the interest of many modern-day fickle fans.

Less-than-exciting all-star games are nothing new to professional sports. The 1968 MLB All-Star game ended in a 1-0 win for the NL, prompting changes to the mound height. The NFL Pro Bowl Game consistently fielded two teams that wanted to escape the weekend without injury, and the play on the field reflected this injury aversion. The 2019 NFL Pro Bowl Game featuring two star-studded rosters had the highest viewership in years, yet it attracted nearly 50% fewer viewers than regular season NFL games in the same year.[31] In 2002, the MLB All-Star Game ended in a 7-7 tie after both teams ran out of pitchers in the 11th inning, making it eerily similar to a Pro Bowl Game. In an effort to generate more excitement, MLB decided that for the next 2 years, the league that won the All-Star Game would have home field advantage in the World Series. That 2-year experiment turned into 14 years.

In a game that has so many statistics attributable to an individual player or team, it seemed counterintuitive to have an advantage (or disadvantage) of a team in the World Series, after 162 regular season games and months of playoffs, codified into the performance of players (many from other teams) in the All-Star game. Over this 14-year period, the American League won 11 of the All-Star Games, but whether or not that truly provided an advantage is debatable. The National League won the 2010, 2011, and 2012 All-Star Games and went on to win the World Series in each of those years. Of the remaining 11 years where the American League won the All-Star Game, the American League won just six World Series. The questionable practice of awarding home field advantage to the league of the winning All-Star team came under scrutiny in 2014 when Cardinals pitcher Adam Wainwright claimed to have thrown Derek Jeter easy pitches in what was Jeter's final All-Star Game. Two years later, home field advantage went back to the team with the better regular season record.

The role of incomplete seasons

The U.S. involvement in World War I led MLB to shorten the 1918 and 1919 seasons from 154 to 140 games, but this marked the first of several shortened seasons of the modern era. Surprisingly, MLB played complete seasons throughout the tumultuous years of World War II, and it wasn't until 1972 that MLB canceled games. The start of the 1972 season was not delayed due to war, but due to something that would plague MLB several times in the latter half of the 21st century: a players' strike.

In a year that saw Roberto Clemente hit his 3,000th hit in his final regular season at bat, the 1972 season also went down in history as the first season to have games canceled since 1919. The regular season began 13 days late due to a players' strike surrounding pension payments and salary arbitration, and MLB made the decision to cancel the 85 games that were scheduled during those 13 days. Teams were disproportionately affected by the cancellations. The canceled games of 1972 had a minimal effect on most teams, with the exception of the Boston Red Sox. The Red Sox finished the season with a record of 85-70. The team that won the American League East, the Detroit Tigers, finished the season with a record of 86-70, a half-game lead over the Red Sox. While the Tigers' success was short-lived, having lost to the Oakland Athletics in the American League Championship, it was little consolation to most Red Sox fans. In the final game of the regular season, the Tigers would defeat the Red Sox 3-1 in what would become the de facto American League East Championship game. Nine years later, MLB would experience its second stoppage due to a players' strike.

The 1981 MLB players' strike over free agent compensation lasted from June 12th to July 31st and canceled 712 games. The strike effectively split the season into two halves. The All-Star Game that was delayed until August 9th would mark the beginning of the second half of the season. In an attempt to avoid inequities resulting from the 1972 strike, MLB restructured the playoffs to have the division winners from the first half of the season play the division winners from the second half of the season. One byproduct of the long mid-season layoff was that all four divisions had different winners in the second half of the season. The revamped playoff structure addressed this disparity but failed in other aspects. For example, the Cincinnati Reds finished the season with a MLB-best record of 66-42, but failed to make the playoffs. The St. Louis Cardinals, who had the second-best record in the National League, also failed to make the playoffs.[32,33] In the American League, the Baltimore Orioles had the same number of overall wins as the New York Yankees with two fewer losses due to an unequal number of games played. Due to the creative playoff structure of 1981, just like the Reds in the National League, the Orioles failed to make the playoffs.

The most significant players' strike impacted the 1994 and 1995 seasons. On August 12, 1994, MLB players would walk off the field in what would become the league's longest strike. The 1994 World Series was canceled. The strike, with its roots in salary caps and revenue sharing, would last through April 2, 1995. Play resumed on April 25, 1995, and the 144-game season of 1995 would end in yet another playoff format. Each league was restructured into three divisions and for the first time, the wild card was introduced, granting a playoff berth for the team in each league with the best record that failed to win their division. The incomplete seasons of 1994 and 1995 would add an asterisk to many statistics from the mid-1990s, but perhaps the most lasting impact was with the fans. The 1995 season saw a 20% decrease in attendance in MLB stadiums. The steroid-fueled home run races of the late-1990s and early-2000s did not generate enough excitement to eclipse the average attendance of 1994. It took well over a decade for fan attendance to reach pre-strike levels.[34]

In the winter of 2020, the world was hit by the coronavirus disease (COVID-19) global pandemic, and the landscape of professional sports would change accordingly. All major professional sports would pause their seasons by mid-March while trying to determine their ability to operate in a COVID-19 environment. Spring training, originally scheduled for late February, would be rebranded as "Summer Camp" as it was delayed until July 1st. From July 23rd through September 27th, teams would play a 60-game regular season schedule. Teams were allowed to carry an expanded pool of players to have the ability to replace roster spots for infected players. The playoffs were expanded to 16 teams, but no games, not even playoff games, would have a live audience. DJ LeMaieu would finish the season batting .364 and Juan Soto would finish the season with an impressive OPS of 1.185. How should these statistics be viewed in light of the circumstances surrounding the 2020 season, in particular only playing 60 games?

Major League Baseball Rule 10.22 (Minimum Standards for Individual Championships) is clear in how individual batting champions are crowned, even in a shortened season. It states that "the individual batting, slugging or on-base percentage champion shall be the player with the highest batting average, slugging percentage or on-base percentage, as the case may be, provided the player is credited with as many or more total appearances at the plate in league championship games, as the number of games scheduled for each club in his club's league that season, multiplied by 3.1 in the case of a Major League player." It also stipulates that any player who does not have the required plate appearances can still qualify if he were charged with the requisite number of plate appearances, essentially changing the denominator, but not the numerator for all batting statistics. This rule has impacted previous shortened seasons with nary an asterisk.[35]

For a player to qualify for most batting statistics leaders in a regular 162-game season, he must have 502 plate appearances. In the shortened season of 1994, Tony Gwynn, the batting champion, achieved an OPS of 1.022. This was the only time Gwynn, who many consider one of the most consistent hitters to ever play the game, achieved an OPS over 1. Gwynn also had the highest batting average of his career that same year, .394. Tony Gwynn's .394 was the highest batting average since Ted Williams hit .406 in 1941 and remains the highest batting average since the 1941 season. How should statisticians view this? In 1994, through no fault of his own and completely attributed to the strike-shortened season, Tony Gwynn had only 475 plate appearances, 48 of those resulting in a base on balls. Most teams in 1994 played only 113 games and using the standard of 3.1 plate appearances per game, a player would qualify for batting statistics with 350 plate appearances. Should any statistics that relate to averages from a shortened season, 1994 and 2020 in particular, be eligible for the record books? Even absent the minimum plate appearance discussion, there is very little debate about Tony Gwynn's batting prowess as he had a batting average over .300 for 19 consecutive seasons from 1983 to 2001.

Shortened seasons can hurt statistics. If Ted Williams's storied 1941 season of batting .406 was cut short by even one day, he would have ended that storied season sub-.400. The last day of that season, the Red Sox had a double header against the Philadelphia Athletics. Going into that final day, Williams had 179 hits on 448 at-bats, putting him just below the elusive .400 average. In the final 2-game homestand of the season, he went 6-for-8 to finish the season at .406. It is easy to say that one shouldn't give in to the hypothetical "what ifs" of seasons past, but there is enough evidence to suggest that the length of the season can have a dramatic impact on how we view the record books.

Shortened seasons are rarely mentioned when discussing Hall of Fame consideration, but they had a potential impact on two players in particular, Larry Walker and Jack Morris. After winning rookie of the year in 1990, Walker began to hit his stride in 1994 with a batting average of .322 and an OPS of .981, the first time he had an OPS over .900. Much of the debate over Walker's improved statistics of the mid-1990s revolved around his trade from Montreal to Colorado in 1995, but in a career where that was plagued by injury, a truncated season can have lasting effects. Playing 10 seasons in a mile-high stadium should not be discounted. From 1995 to 2004, Walker had an OPS of 1.176 at Coors Field compared to .905 when he was playing in away ballparks. Additionally, he had a .383 batting average and .713 slugging percentage at Coors Field, compared to a .281 batting average and .518 slugging percentage away from home. The altitude advantage should not be ignored, but shortened seasons had an impact on Walker's career hits and home runs. Often scrutinized, Larry Walker had 383 career home runs and 2,160 hits. In his final year of eligibility, Larry Walker was inducted into the Hall of Fame after receiving 76.6% of the votes from the Baseball Writers Association of America in 2020, just over the required 75%. Not all razor thin margins have a happy ending.

Jack Morris's 18-year career as a pitcher from 1977 to 1994 resulted in five All-Star appearances, four World Series Championships, a World Series MVP award, and 254 wins and three 20-win seasons. Jack Morris was one of the most reliable, consistent pitchers in the history of the game. In 527 starts (515 of which were consecutive), Morris threw 175 complete games. What turned out to be his final year in MLB, 1994, was cut short due to the players' strike and Morris made his final appearance on August 9, 2014. After a failed comeback during spring training for the Cincinnati Reds in 1995, Jack Morris retired. Jack Morris was on a trajectory to make the Hall of Fame after garnering 67.7% of the votes in his second-to-last year of eligibility. However, in 2014, Morris's final year of eligibility, he received 61.5% of the vote, largely due to two prominent pitchers who were first-ballot selectees, Greg Maddux and Tom Glavine. Had Jack Morris pitched one additional season, his Hall of Fame eligibility would have extended into the 2015 voting year and his share of the votes would have likely increased, potentially over the 75% threshold. A counterargument is that pitchers Randy Johnson, Pedro Martinez, and John Smoltz were all elected to the Hall of Fame in 2015. In the end, Jack Morris found his way into the Hall of Fame after the Modern Era Committee selected him for enshrinement in 2018, but an argument can be made that the strike of 1994 had an impact on Morris's unconventional, lengthy path to Cooperstown.[36]

The final out

Statistics need context and must be viewed through the lens of the time-frame and environment in which they are recorded. Every time period in the modern era of baseball has circumstances that make it difficult to compare players across generations. Scandals, rule changes, racism, incomplete seasons, PEDs, and the unethical use of technology rightfully cast a shadow on how we view player performance. The asterisks associated with each time period create cognitive dissonance and generate debates in which there are very few right and wrong answers.

It has been said that money is the root of all evil. Even if it isn't the root of *all* evil, it is at the root of many of the problems that professional baseball has faced over the past hundred years. An example of this is baseball's love-hate relationship with gambling. In 1919, eight players of the Chicago White Sox threw the World Series in an attempt to profit from a gambling ring. Their trial in 1921 led to a not guilty verdict, but all eight received a lifetime ban from professional baseball. The desire to make money, while unethical, was explainable given the lack of monetary benefit that most players faced in the early 1900s. In 1919, Ty Cobb was the highest paid player in professional baseball, earning $20,000 per year. Over 100 years later, Cobb's salary is still only the equivalent of about $300,000, or over $250,000 below the league minimum salary. Fast-forward nearly 100 years, and we find that MLB has embraced the role of gambling in professional baseball. On May 14, 2018, the United States Supreme Court found the Professional and Amateur Sports Protection Act (PASPA), the federal law prohibiting states from authorizing sports betting, to be unconstitutional. Within 18 months, sports betting was legalized in 21 states and the District of Columbia. The American Gaming Association estimates MLB stands to gain over $1 billion in annual revenue as a result of legal gambling. The majority of the revenue increase is estimated to come through fan engagement in the form of media rights, sponsorship, merchandise, and ticket sales.[37] Although players are prohibited from betting on professional baseball, in 100 years MLB has embraced a practice that was part of one of its most shameful periods.

The current level of player salaries, with the league minimum over $500,000, reduces the temptation for players to fix games, but has caused a high-stakes game of whack-a-mole for MLB. Player salaries and collective bargaining agreements caused incomplete seasons in the 1970s, 1980s, and 1990s. As player salaries increased, baseball experienced the emergence of PEDs. The expanded drug testing in the early 2000s helped level the playing field in terms of player performance, but teams found ways to gain a competitive advantage through the unethical use of technology. Given recent controversy of privileged communication, perhaps the role of sabermetrics is expanding beyond player evaluation and in the direction of anomaly detection in order to maintain integrity in the game. As MLB seeks answers to legal, technological, and ethical questions that will certainly arise over the next 100 years, there will be new debates and criteria for player evaluation. The answers are buried in the numbers.

While statistics can be relative to the time period in which they were recorded, sabermetrics author Gabe Costa once famously said, "if you develop an algorithm to determine the best hitter of all time and it gives you anyone other than Babe Ruth, there is a flaw in your logic." Maybe it is that easy.

References

1. "digit, n." OED Online, Oxford University Press, December 2020, www.oed.com/viewdictionaryentry/Entry/52611. Accessed June 30, 2020.
2. Ore, Oystein (1948). *Number Theory and its History*. McGraw-Hill Book Company, Inc.
3. Costa, Gabriel (2016). "MLB Network Presents – 56: The Streak." https://www.mlb.com/video/56-probability-c702744983. Accessed July 8, 2020.
4. Pesca, Mike (2009). "The Man Who Made Baseball's Box Score a Hit." https://www.npr.org/templates/story/story.php?storyId=106891539. Accessed May 4, 2020.
5. Schiff, A. J. (2008). *'The Father of Baseball': A Biography of Henry Chadwick*. McFarland and Company, Inc., p.59–89.
6. Chadwick, Henry (1872). *Beadle's Dime Baseball Player*. Beadle and Company.
7. Schwarz, Alan (2005). *The Numbers Game: Baseball's Lifelong Fascination with Statistics*. St. Martin's Griffin, p.10–11.
8. Topps. "75 Years as a Proud American Icon." https://www.topps.com/history. Accessed January 7, 2021.
9. MLB Glossary. "Advanced Statistics." http://m.mlb.com/glossary/advanced-stats. Accessed September 3, 2020.
10. Athiviraham A, Bartsch A, Mageswaran P, et al. (2012). "Analysis of Baseball-to-Helmet Impacts in Major League Baseball." *The American Journal of Sports Medicine* 40(12), p.2808–2814.
11. Sowell, M. (1991). *The Pitch That Killed*. Collier/Macmillan.
12. MLB Glossary. "Strike Zone." http://m.mlb.com/glossary/rules/strike-zone. Accessed September 3, 2020.
13. Lewis, Michael (2004). *Moneyball: The Art of Winning an Unfair Game*. WW Norton & Company, p. 2–30.
14. McKenna, Brian. "Bob Ferguson." https://sabr.org/bioproj/person/bob-ferguson-2. Accessed January 14, 2021.
15. Paine, Neil (2016). "Why Baseball Revived A 60-Year-Old Strategy Designed to Stop Ted Williams." https://fivethirtyeight.com/features/

ahead-of-their-time-why-baseball-revived-a-60-year-old-strategy-designed-to-stop-ted-williams/. Accessed January 15, 2021.

16. Noga, Joe (2020). "Aligned Right: Major League Baseball's Infield Shift Trend has Roots in Cleveland." https://www.cleveland.com/tribe/2020/02/aligned-right-major-league-baseballs-infield-shifting-trend-has-roots-in-cleveland.html. Accessed January 15, 2021.

17. Simon, Mark (2018). "Has the Shift Seen its Day?" 2018 SABR Analytics Conference. Available at https://sabr.box.com/shared/static/3t158zqgoia0z6u6sua17q1nm3k2nab9.ppt.

18. Carlton, Russell (2018). "Baseball Therapy: How to Beat the Shift." https://www.baseballprospectus.com/news/article/40088/baseball-therapy-how-beat-shift/. Accessed November 12, 2020.

19. Abrams, Daniel M. and Mark J. Panaggio (2012). "A Model Balancing Cooperation and Competition can Explain our Right-handed World and the Dominance of Left-handed Athletes. *Journal of the Royal Society Interface*, p. 2718–2722.

20. Leeke, Jim (2017). *From the Dugouts to the Trenches: Baseball During the Great War.* University of Nebraska Press, p. 192.

21. Weintraub, Robert (2013). "Two Who Did Not Return." https://www.nytimes.com/2013/05/26/sports/baseball/remembering-the-major-leaguers-who-died-in-world-war-ii.html. Accessed May 4, 2020.

22. Schwarz, Alan (2005). The Numbers Game: Baseball's Lifelong Fascination with Statistics. St. Martin's Griffin, p.16.

23. Chass, Murray (2002). "Agreement Reached in Baseball Contract Negotiations." https://www.nytimes.com/2002/08/30/sports/agreement-reached-in-baseball-contract-negotiations.html. Accessed December 15, 2020.

24. Mitchell, George (2007). *Report to the Commissioner of Baseball of an Independent Investigation into the Illegal Use of Steroids and Other Performance Enhancing Substances by Players in Professional Baseball.* Available at https://files.mlb.com/mitchrpt.pdf.

25. Office of the Commissioner of Baseball (2019). "MLB Rule 5.07: Pitching." Available at https://content.mlb.com/documents/2/2/4/305750224/2019_Official_Baseball_Rules_FINAL_.pdf.

26. Imber, Gil (2015). "Reviewing Instant Replay: Observations and Implications from Replay's Inaugural Season." *The Baseball Research Journal*, Spring 2015. Available at https://sabr.org/journals/spring-2015-baseball-research-journal/.

27. MLB Glossary. "Statcast." http://m.mlb.com/glossary/statcast/. Accessed November 5, 2020.

28. Baseball Savant "MLB Instant Replay Database." https://baseballsavant.mlb.com/replay. Accessed November 10, 2020.

29. Costa, Gabe (2013). "By the Numbers: The World Series and the 'Footballization' of Baseball." https://newyork.cbslocal.com/2013/10/08/by-the-numbers-the-world-series-and-the-footballization-of-baseball/. Accessed August 12, 2020.

30. MLB Official Info. "MLB Miscellany: Rules, Regulations and Statistics." http://mlb.mlb.com/mlb/official_info/about_mlb/rules_regulations.jsp. Accessed August 2, 2020.

31. Florio, Mike (2018). "Pro Bowl Ratings up with a Caveat." https://profootball-talk.nbcsports.com/2018/01/29/pro-bowl-ratings-up-with-a-caveat/. Accessed August 22, 2020.

32. Calcaterra, Craig (2020). "Looking Back at Baseball's Previously Shortened Seasons." https://mlb.nbcsports.com/2020/03/13/looking-back-at-baseballs-previously-shortened-seasons/. Accessed November 7, 2020.

33. Bambuca, Chris (2020). "Explaining the 1981 MLB Season: How Baseball Survived Shortened Year." https://www.usatoday.com/story/sports/mlb/2020/03/15/1981-mlb-season-coronavirus-delay-baseball/5054780002/. Accessed January 10, 2021.

34. The Baseball Cube. "MLB Attendance History." http://www.thebaseballcube.com/topics/attendance/. Accessed January 5, 2021.

35. Office of the Commissioner of Baseball (2012). "MLB and MLBPA Reach Agreement on Rule 10.22(a)." https://www.mlb.com/news/mlb-and-mlbpa-reach-agreement-on-rule-1022a/c-38781290. Accessed May 30, 2020.

36. Piling, Rich. National Baseball Hall of Fame and Museum. "Jack Morris." https://baseballhall.org/hall-of-famers/morris-jack. Accessed January 5, 2021.

37. American Gaming Association (2020). "How Much do Leagues Stand to Gain from Legal Sports Betting?" https://www.americangaming.org/wp-content/uploads/2018/10/Nielsen-Research-All-4-Leagues-FINAL.pdf. Accessed January 15, 2021.

Epilogue

10

Where have we been? Where do we go from here? A final word from the editor

For those of us who were weaned on baseball, the game will forever be young. The smell of the grass, the crack of the bat—even the "choosing up of sides" in the sandlots—these endure throughout life. Although we all age and our knees creak with arthritis and our eyesight is dimmed, the memories remain unimpaired.

Numbers are important. Because they are the only quantitative way to compare one ball player to another, and, indeed, one era to another. It is true that we look back in horror at some of the great scandals, such as the 1919 World Series and lament the events which soiled that Fall Classic. What career numbers would Joe Jackson have posted? Heroes like Jackie Robinson and Larry Doby bravely faced discrimination every day of their Hall of Fame careers, while previous stars such as Josh Gibson and Buck Leonard had no chance to make the big time. What would their numbers have looked like? Alas, history cannot be changed.

On the other hand, how many numbers have been skewed? When sluggers' bodies balloon up and cumulative averages increase (like home run percentages), the purity of the game is called into question… and the plausibility of comparisons and contrasts is compromised…

Yet, as water seeks its own level, the fans—ever resilient—take it in stride… and, after a while, all's well again…

Electronics and robotics will undoubtedly make even more of an impact in the future than they already have. Perhaps the day will come when umpires are not required, at least not on the field… (I hope not!)

For the Major Leagues in the United States (and Canada's Blue Jays), it seems inevitable that universal expansion will become a reality. Cuba, the Dominican Republic, Puerto Rico, and Panama, among other nations, have provided many stars over the past three generations. Certainly, there have also been a great many Asian players. If the game truly catches on in Europe, the idea of a literal *World-Wide Series* could be realized.

Economics is another factor. Alex Rodriguez has earned nearly a half-billion dollars in salary for his career. Can a billionaire player be far behind?

Sabermetrics. DOI: http://dx.doi.org/10.1016/B978-0-12-822345-1.00010-6

There does not seem to be any tapering off of the rising of revenues, at least for the foreseeable future. Will baseball outprice itself?

In the long run, I don't think so. Once the spring comes... one can literally smell baseball in the air...

In the end, baseball has been a gift... given to us by our Creator. So great is the game that, in the purest sense, it becomes part of our souls.

I hope you enjoyed our book. I believe our contributors were magnificent.

I leave all that is left to say about the game to the late Jim Bouton. The Bulldog, who was both a pitcher and an author, put it as well as anybody:

> *"A ballplayer spends a good piece of his life gripping a baseball, and in the end it turns out that it was the other way around all the time."*

Index